Springer Series in
CHEMICAL PHYSICS 76

Springer
Berlin
Heidelberg
New York
Hong Kong
London
Milan
Paris
Tokyo

Springer Series in
CHEMICAL PHYSICS

The purpose of this series is to provide comprehensive up-to-date monographs in both well established disciplines and emerging research areas within the broad fields of chemical physics and physical chemistry. The books deal with both fundamental science and applications, and may have either a theoretical or an experimental emphasis. They are aimed primarily at researchers and graduate students in chemical physics and related fields.

63 **Ultrafast Phenomena XI**
Editors: T. Elsaesser, J.G. Fujimoto, D.A. Wiersma, and W. Zinth

64 **Asymptotic Methods in Quantum Mechanics**
Application to Atoms, Molecules and Nuclei
By S.H. Patil and K.T. Tang

65 **Fluorescence Correlation Spectroscopy**
Theory and Applications
Editors: R. Rigler and E.S. Elson

66 **Ultrafast Phenomena XII**
Editors: T. Elsaesser, S. Mukamel, M.M. Murnane, and N.F. Scherer

67 **Single Molecule Spectroscopy**
Nobel Conference Lectures
Editors: R. Rigler, M. Orrit, T. Basché

68 **Nonequilibrium Nondissipative Thermodynamics**
With Application to Low-Pressure Diamond Synthesis
By J.-T. Wang

69 **Selective Spectroscopy of Single Molecules**
By I.S. Osad'ko

70 **Chemistry of Nanomolecular Systems**
Towards the Realization of Molecular Devices
Editors: T. Nakamura, T. Matsumoto, H. Tada, K.-I. Sugiura

71 **Ultrafast PhenomenaXIII**
Editors: D. Miller, M.M. Murnane, N.R. Scherer, and A.M. Weiner

72 **Physical Chemistry of Polymer Rheology**
By J. Furukawa

73 **Organometallic Conjugation**
Structures, Reactions and Functions of d–d and d–π Conjugated Systems
Editors: A. Nakamura, N. Ueyama, and K. Yamaguchi

74 **Surface and Interface Analysis**
An Electrochmists Toolbox
By R. Holze

75 **Basic Principles in Applied Catalysis**
By M. Baerns

76 **The Chemical Bond**
A Fundamental Quantum-Mechanical Picture
By T. Shida

Series homepage – springeronline.com

Volumes 1–62 are listed at the end of the book

Tadamasa Shida

The Chemical Bond

A Fundamental Quantum-Mechanical Picture

With 40 Figures

Springer

Professor Tadamasa Shida
Kanagawa Institute of Technology
Atsugi 243-0292, Japan
E-mail: shida@kyoto.email.ne.jp

Series Editors:
Professor A. W. Castleman, Jr.
Department of Chemistry, The Pennsylvania State University
152 Davey Laboratory, University Park, PA 16802, USA

Professor J.P. Toennies
Max-Planck-Institut für Strömungsforschung
Bunsenstrasse 10, 37073 Göttingen, Germany

Professor W. Zinth
Universität München, Institut für Medizinische Optik
Öttingerstr. 67, 80538 München, Germany

KAGAKU KETSUGO by Tamasada Shida

Originally published in Japanese in 2001 by Iwanami Shoten, Publishers, Tokyo. This english language edition publisehd 2004 by Springer-Verlag, Heidelberg by arrangement with the author through Iwanami Shoten, Publishers, Tokyo

ISSN 0172-6218

ISBN 3-540-20638-8 Springer-Verlag Berlin Heidelberg New York

Library of Congress Cataloging-in-Publication Data

Springer-Verlag is a part of Springer Science+Business Media.

springeronline.com

Typesetting: Frank Herweg, Leutershausen
Cover concept: eStudio Calamar Steinen
Cover production: *design & production* GmbH, Heidelberg

Printed on acid-free paper 57/3141/ts - 5 4 3 2 1 0

Preface to the Original Japanese Version

Man has created a branch of natural science, called chemistry, which through many centuries has been concerned with the separating and purifying of matter and the measuring of the changes in mass and volume which accompany chemical reactions. However, it is just a century ago that man finally realized that matter consists of invisible entities, i.e., atoms and molecules. Since this discovery, it became necessary for man to comprehend matter, not in terms of empirical knowledge such as the periodicity of elements, but in terms of concepts which acknowledge the quantum nature of matter, such as the atomic nucleus and electronic shell structure, all of which are imperceptible.

For the rudimentary understanding of atoms and molecules it has become mandatory to grasp the hitherto unknown field of quantum mechanics. This means that it is a necessity to learn a new language, so to speak, to understand this insensible microscopic cosmos, which cannot be discussed in the familiar language used to describe the macroscopic world. This dramatic change and the necessity for new terminology is one reason that chemistry appears difficult to understand on an atomic/molecular level. This situation differs from that in physics where the concepts of force, energy, and electromagnetism are easily associated with such familiar things in daily life as hydroelectricity, motors, earthquakes, and so forth.

Let me give a concrete example of the necessity of understanding quantum mechanics: why do two electrically neutral hydrogen atoms produce a stable hydrogen molecule by simply approaching each other? Likewise, why are two repelling protons subordinated to the binding power of just a single electron to give a thermodynamically stable hydrogen molecular ion H_2^+? In order to answer these questions there is no way around using a quantum mechanical explanation.

Quantum mechanics has made rapid progress since its birth, in the last three quarters of the last century, to the current level where most problems of atoms and molecules are now solved. This is in contrast to the history of traditional chemistry which advanced in small, time-taking steps. Fortunately, however, chemists have been eager to understand and apply the essence of quantum mechanics since its very beginning to foster a field called quantum chemistry. Behind the successful application of this new field was a great deal of ground-leveling effort to accept this new science, i.e., the accumulation and compilation of knowledge based on a vast amount of experimental fact and intuitive inference within the framework of pre-quantum theory. G.N. Lewis (1875–1946) and I. Langmuir (1881–1957) among many others are such forerunners. The Nobel prize laureates who introduced quantum theory into chemistry include L.C. Pauling (1954), R.S. Mulliken (1966), G. Herzberg (1971), K. Fukui (1981), R. Hoffmann (1981), and J.A. Pople (1998). The fundamental understanding of chemical bonds, the theme of this book, owes very much to these pioneers.

Nowadays, one can obtain the results of quantum chemical computations rather easily thanks to the aforementioned pioneers and to the remarkable progress in computers. However, I feel that the wide use of computers may not necessarily engender a fundamental understanding of the true nature of chemical bonding. Suppose that someone gets the results of computations via computer of the hydrogen molecule without knowing the true implication of such a sentence as "the essential stability of the ground state hydrogen molecule is due to the exchange force which is inexplicable in terms of classical mechanics and electrodynamics", then, that computation will be of little meaning.

Broadly speaking, this book is aimed at readers who have finished a beginning course in quantum chemistry and wish to advance a step further. In hopes of assisting understanding, I have incorporated various experimental and historical examples which are of a practical nature since chemistry is essentially a science of everyday life and nature, based on experiences.

In Chap. 1, I will give a brief look at the electron in a chemical environment. Chapter 2 is devoted to a summary of the minimum essentials of quantum mechanics necessary for the understanding of chemical bonding. I presume that the readers will complement the discussion in this chapter by referring to any fundamental textbook on quantum mechanics. On the basis of Chap. 2, I will discuss one-electron atoms in Chap. 3. The role of this chapter is two-fold: it provides a fundamental basis for Chap. 4 and future chapters, it provides the readers with practical examples utilizing the abstract concepts from Chap. 2. I will deliberately reiterate the contents of Chap. 2 to make the readers familiar with them. In Chap. 4, I will explain the multi-electron atom as a building block in a variety of molecules. Chapter 5 will cover the transition from one-atom systems to many-atom systems, i.e., molecules. After that, especially in Chap. 6, I will explore several concepts on bond formation and the features of some typical chemical bonds, up through Chap. 8.

As for electron spin, which is a real quantum mechanical phenomenon, I have presented comprehensive explanation, the reason being that electron spin is as important as electron energy, and is usually quite elusive for those at an introductory level.

The understanding of multi-electron atoms is indispensable for the understanding of molecules. The theory of multi-electron atoms progressed very rapidly in its early days, which is proven by the fact that such textbooks as *Atomic Physics* by M. Born, first published in 1935, and *The Theory of Atomic Spectra* by E.U. Condon and G.H. Shortley, also published in 1935, are still being printed. Besides these classics, there are plenty of textbooks on multi-electron atoms so I have restricted my discussion on this subject to the minimum level essential for the understanding of chemical bonding. On the other hand, I have devoted quite a few pages in Chaps. 6 and 7 to the hydrogen molecular system. This is because I consider that the essence of the chemical bond formation is in these systems. As a result of this emphasis on the hydrogen molecular systems, many of the important subjects related to general diatomic molecules had to be omitted. A partial remedy was made in the discussion on hydrogen fluoride in Chap. 8, however. Fortunately, 'The Bible' on diatomic molecules by G. Herzberg is still commercially available (cf. Refs. for Chap. 6), among many other textbooks, so that readers can refer to any of these books for detailed discussion. In the final chapter, Chap. 8, I have presented semiquantitative or qualitative interpretations of the results of molecular orbital analyses on some polyatomic molecules with the hope that the reader may gain a new insight into the chemical bond.

Today, computer analysis based on the molecular orbital and the density functional theories is very popular. However, since I am not an active developer of computational methodology but rather a passive user of the available computer programs, I have chosen to place emphasis on the introduction of the underlying theory essential to the clear understanding of chemical bonds and other basic concepts.

Recently, the collegiate curricula in physics and chemistry is confronted by serious problems owing to the (Japanese) government's policy of simplification of education in high schools. This problem, as it manifests itself in such a fundamental field as quantum chemistry, will become ever the more serious. Many former textbooks used for undergraduates are now being regarded as graduate-level texts. Because of this trend, however, clear and well-balanced textbooks are being published. Compared to these recent books, the present book may be a little hard to crack for undergraduates partly because of the limitations of space and partly because of the lack of my own ability in presenting concepts in a simplified manner. However, it is my view that those learners who are motivated and interested sometimes want to have something hard to chew on and are repelled by anything unchallenging. I am not sure that the present book will be accepted as such, but at any rate, I have aimed and endeavored to prepare this book with a belief that patient reading leads to eventual understanding. The reader may notice that I redundantly present information and explanations. This is a deliberate attempt to assist the reader's 'chewing of the cud'.

It is proper to cite every reference individually; this rule was not obeyed, however, due to the pre-established style of this series by the publishing company. However, I have incorporated some of the original sources in the text when I felt it appropriate. The asterisk implies that the level may be a little above the general level of the book.

Professor Suehiro Iwata at National Institution for Academic Degrees, with whom I have collaborated on many occasions since our days at the Institute of Physical and Chemical Research, reviewed the manuscript and added proper sentences here and there. Professor Michio Matsushita, now at Tokyo Institute of Technology, made a number of useful comments during his busy stay at Leiden University. Professor Taichi Kugo at Physics Department of Kyoto University advised me on the description of nuclear physics in Chap. 6. I thank Professor Takamasa Momose at Kyoto University, with whom I had collaborated for many years, for his general commentary.

Kyoto, August 2001 *Tadamasa Shida*

Preface to the Translated Version

This book was originally published in Japanese by Iwanami Shoten, Publishers in 2001 as one of eighteen serial volumes. The series is divided into three levels: elementary, intermediate, and advanced. This book is classified as intermediary and, as stated in the preface for the original publication, targets senior undergraduate students. However, as is also mentioned in the preface, there are sections of the book, e.g., Chap. 3.4, which may be a bit beyond the undergraduate level.

In preparing the English version I have followed the original as closely as possible except for a few changes and corrections. I would like to acknowledge critical comments of Professors Hirohiko Kono and Kenji Ohmori at Tohoku University. I also thank my former students Dr. Mizuho Fushitani at Frei Universität Berlin and Dr. Hiroyuki Katsuki at the University of Zurich for their comments on my description in Chapter 6.

The English manuscript was first edited by Assistant Professor Chris A. Bowen, my present colleague at Kanagawa Institute of Technology. After Prof. Bowen's revisions to the translation, Dr. Jon T. Hougen at NIST kindly agreed to look over the scientific content and presentation of the completed manuscript, and I am grateful for his many comments also. Professor Takeshi Oka at the University of Chicago kindly pointed out some historical inaccuracies. Dr. Hiromichi Hoshina who moved recently to the University of Southern California and Mr. Yoshiyasu Kato and Mr. Susumu Kuma, now doctral candidates under the guidance of Professor Momose, offered generous assistance with computer-editing when I had need of it.

In order to update the references some recent publications are added in the English version.

Kyoto, August 2003 *Tadamasa Shida*

Contents

1 The Electron: A Primadonna in Chemical Bonding

Molecules are an assemblage of positively charged nuclei and negatively charged electrons. The geometrical arrangement of nuclei, the mass of which is much larger than that of the electron, is nearly fixed under normal conditions, while the electrons are distributed around the nuclei. Thus, with a nucleus regarded as one center, a molecule can be viewed as a multi-centered, multi-electron system. The theory of chemical bonding is concerned with the problem of how molecules are formed. In chemical bonding, the electron plays the role of primadonna.

Nuclei consist of protons and neutrons. Previously, these two classes of particles were regarded as elementary particles, and together with the electron, were considered the basic building blocks of matter. Nowadays, it is known that the proton and neutron are not single particles but consist of more fundamental particles, called quarks. The electron remains an elementary particle as before. Furthermore, the electron is unique in that it is the lightest charged and stable elementary particle. In chemistry, however, one can safely regard the proton and neutron as stable elementary particles like the electron because the energy necessary to dissociate the proton and neutron into quarks is orders of magnitude larger than the chemical energy.

The variations in the properties of matter can be basically attributed to the fact that the electron is so light and negatively charged, such that it can interact electrostatically in various ways with positively charged nuclei and other negatively charged electrons. In chemical events, the electron interacts with both nuclei and other electrons. I will briefly introduce the behavior of the electron in a certain simple environment, and then, give a short summary of the history of the electron from a quantum mechanics viewpoint.

1.1 The Electrostatic Potential Felt by the Electron is Reflected in the Color of Matter
The energy state of the electron is determined by the surrounding electrostatic potential. The difference in energy matches the energy of photons absorbed or emitted by matter.

1.2 The Electron in a Square Well Potential
In order to confirm the statement of Sect. 1.1, I will give an example of the one-dimensional square well potential where the energy and the eigenfunction of the electron can be exactly solved.

1.3 The Electron Spin: as Important as the Energy
When the electron moves in space, it possesses orbital angular momentum in accordance with classical electrodynamics. In addition, there is another angular momentum component which has no classical analogue; this is the intrinsic angular momentum called the electron spin.

1.4 A Brief History of the Quantum Mechanical Description of the Electron
After the introduction of preliminaries by Bohr, Schrödinger constructed non-relativistic quantum mechanics.

1.1 The Electrostatic Potential Felt by the Electron is Reflected in the Color of Matter

Let us think of an electron in liquid ammonia. Liquid ammonia is easily obtained as a colorless fluid when gaseous ammonia at STP is cooled below the boiling point (−33.4°C) at atmospheric pressure. Introduction of a piece of alkali metal to the liquid turns it to a blue-ink colored solution. What happens is the release of an electron from the alkali metal, leaving a solvated electron and a positive alkaline ion. The electron enjoys a comfortable environment by electrically polarizing the surrounding ammonia molecules. The electron, thus stabilized, can be excited by absorbing red to yellow components of white light. As a result, the solution turns to blue, the complementary color to the absorbed components.

Figure 1.1(a) shows a schematic illustration of the electron polarizing the surrounding ammonia molecules. Figure 1.1(b) is a qualitative sketch of the electrostatic potential as a function of the distance of the electron from the origin. According to quantum mechanics, whenever the electron is confined

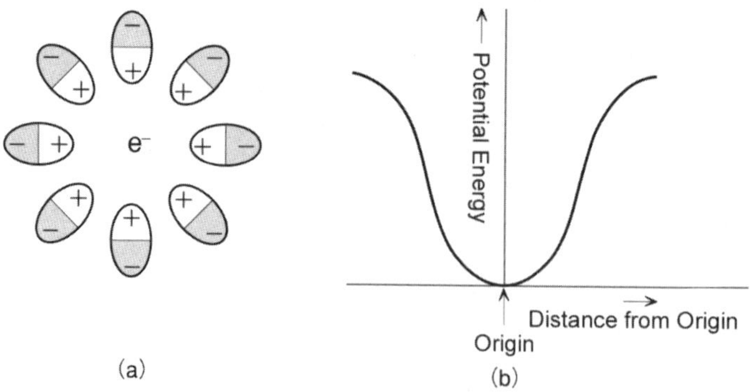

Fig. 1.1. (**a**) Schematic sketch of an electron e^- in liquid ammonia. (**b**) A qualitative description of the potential felt by the electron. In (**a**), the ammonia molecules are supposed to be partially polarized by the central electron

in such a local potential, the energy of the electron is restricted to certain discrete values which are determined by the shape of the potential.

According to M.K.E.L. Planck, the energy of light is given by the product of a constant $h \cong 6.6 \times 10^{-34}$ J s (now known as the Planck constant) and the frequency ν (whose unit is reciprocal of seconds, $\mathrm{s}^{-1} = \mathrm{Hz}$). When the difference of energies allowed for an electron in an electrostatic potential satisfies the equation, $\Delta E = h\nu$, the electron absorbs the energy of light.

Unfortunately, the exact shape of the potential shown schematically in Fig. 1.1(b) is not known. Therefore, exact energies of the electron shown cannot be determined, nor can the difference ΔE, so that one cannot predict precisely the frequency of light to be absorbed by the solution.

In contrast, the electrostatic potential for an electron in a hydrogen atom, to be dealt with in Chap. 3, is uniquely given as a function inversely proportional to the distance of the electron from the proton (Coulombic potential). In this case, the energy states of the electron are known exactly, which discloses all the information on the electron in the atom. However, as will be seen in Chap. 3, the exact solution is obtained only after a lengthy mathematical calculation. Therefore, the conclusion stated in the title of Sect. 1.1 is reached only after a long excursion in the case of the Coulombic potential.

However, in order to demonstrate that the statement in the title holds good in general, including the case of an electron in liquid ammonia, I propose to replace the realistic but indeterminable potential in Fig. 1.1(b) with an approximate but determinable potential.

For this purpose, I regard the electron as confined by the blocking ammonia molecules in all three dimensions, front and rear, left and right, and up and down. Thus, this becomes a three dimensional problem. For the sake of simplicity, however, I will reduce the dimensions from three to one, taking x as a single variable. Furthermore, I assume that the potential felt by the electron can be approximated by the potential in Fig. 1.2(a).

The assumed potential is such that the potential function V is equal to 0 for the range of $|x| < d$, and is infinity for $|x| > d$, where d stands for the distance from the origin to the barriers of infinite potential. Of course, the real potential will be finite, but assuming infinity allows the derivation of elementary mathematical relations and solutions, as will be shown in Sect. 1.2. The potential in Fig. 1.2(a) is called the square well potential.

1.2 The Electron in a Square Well Potential

When an electron of mass equal to m_e is confined in the region of $|x| < d$ as in Fig. 1.2(a), the classical expression of the kinetic energy is $p^2/(2m_\mathrm{e})$, where p stands for the momentum of the electron. Since the potential energy in this region is assumed to be 0, the total energy also is equal to $p^2/(2m_\mathrm{e})$. As will be shown in Chap. 2 and on, to solve the problem quantum mechanically for the energy state of the electron is tantamount to solving the Schrödinger

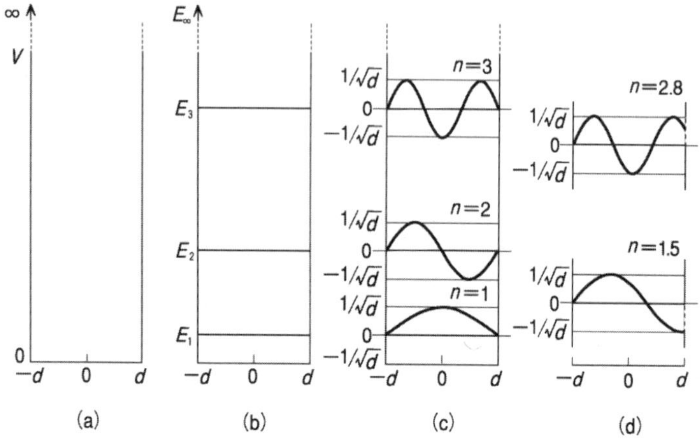

Fig. 1.2. (**a**) A square well potential. (**b**) The energy levels of an electron in the potential. (**c**) The energy functions of the allowed states in the case of n being integers. (**d**) Unallowable states for cases of n being non-integers

equation in (1.1) to obtain the function $\phi(x)$, which describes the state of the electron, and to get the energy value E in that state.

$$\hat{H}\phi(x) = E\phi(x) \tag{1.1}$$

The symbol $\hat{H}$ in (1.1) is called the Hamilton operator or Hamiltonian, and stands for a mathematical manipulation. While a general explanation of such manipulations will be given in Chap. 2, for this simple example, we can take $\hat{H}$ to be equal to a constant times the second derivative of $\phi(x)$ with respect to x, i.e., $\hat{H} = -\hbar^2/(2m_e) \times (\mathrm{d}^2/\mathrm{d}x^2)$. Here, the symbol $\hbar$ means $\hbar \equiv h/(2\pi)$, and the caret is to emphasize that $\hat{H}$ is an operator. In quantum mechanics, all physical quantities such as the energy are associated with corresponding operators. It is an implicit understanding that there is always a function, next to the operator, to be operated on by the operator.

The physical significance of the function $\phi(x)$ will be fully described in Sect. 1.4.2 as well as in Chap. 2. For the moment, it suffices to say that the square of the absolute value of $\phi(x)$ is proportional to the probability of finding the electron in the range between x and $x + \mathrm{d}x$.

Summarizing the constants $\hbar$ and m_e in $\hat{H}$ and rewriting (1.1), one obtains the equation:

$$\frac{\mathrm{d}^2}{\mathrm{d}x^2}\phi(x) = \left(-\frac{2m_eE}{\hbar^2}\phi(x)\right) = \text{const} \times \phi(x) \tag{1.2}$$

The left-hand side represents the second derivative of $\phi(x)$ with respect to x, while the right-hand side demands that the result of the derivative be a simple multiple of the original function $\phi(x)$. Since this is a special demand, the function $\phi(x)$ must be restricted accordingly. As a trial, let us take $\phi(x)$ as equal to a sine function and put $\phi(x) = \sin k(x+d)$. Then, (1.2) turns out to be (1.3), which satisfies the special relations demanded.

$$\frac{\mathrm{d}^2}{\mathrm{d}x^2}\phi(x) = -k^2 \sin\{k(x+d)\} = -k^2\phi(x) \tag{1.3}$$

Such equations as (1.2) and (1.3) which have the form of operator × function = const × function are classified as an eigenvalue equation in mathematics, where the symbol × signifies that the function is affected by the operation of the preceding operator or the constant. A function satisfying such a relation is called an eigenfunction, and the accompanying constant is the eigenvalue. In physics and chemistry some solutions to these equations are disregarded for physical reasons, even though they are mathematically correct. For examples of such solutions, one can consider a function which diverges infinitely or becomes abruptly discontinuous at some point, a function which does not connect two merging regions smoothly, or a function which gives plural values of the function for a certain set of variables. The conditions set to remove these nonphysical solutions are called boundary conditions.

In the present case of the square well potential, the function $\phi(x)$ must vanish at $x = -d$ because of the infinite potential. Thus, one must require the condition $\phi(x = -d) = 0$. This condition is automatically fulfilled by the trial sine function. For the same reason, one has to require the condition $\phi(x = +d) = \sin(2kd) = 0$.

This latter requirement is equivalent to the enforcement of $2kd = n\pi$ with $n = 0, \pm1, \pm2, \ldots$. Thus, the function satisfying the two conditions above can be written as below.

$$\phi_n(x) = \sin\{n\pi(x+d)/(2d)\} \tag{1.4}$$

Substitution of the above function to (1.2) yields the relation $(n\pi/2d)^2 = 2m_\mathrm{e}E/\hbar^2$, from which it is found that the allowed energies of the electron are restricted to the following discrete values:

$$E_n = n^2h^2/(32m_\mathrm{e}d^2) \tag{1.5}$$

The horizontal lines in Fig. 1.2(b) represent schematically the discrete values of energy measured from the bottom of the potential well. The sinusoidal curves in Fig. 1.2(c) correspond to the sine functions in (1.4), each of them representing the state function of the electron at the respective energy level. The physical meaning of the state function will become clearer in due course. At the moment, it is sufficient to accept that the ordinate of Fig. 1.2(c) represents the amplitude of the state functions, the dimension of which is the reciprocal of the square root of length (cf. Sects. 2.1.1 and 2.2.1).

According to (1.4), the function $\phi_n(x)$ in the state with $n = 0$ becomes 0 for all x, which means that the electron cannot be in that state. In the states with $n = \pm 1, \pm 2, \pm 3, \ldots$, the sign of the functions $\phi(x)$ reverses according to the sign of n. The sign of the function itself, however, seldom has a physical meaning, as will be elaborated in Chap. 2. The quantity of physical significance associated with the function is the square of the absolute value of the function. Thus, one can ignore the difference in the sign of the functions in Fig. 1.2(c). The figure depicts the case for $n = +1, +2$, and $+3$. Since the state corresponding to $n = 0$ does not exist, the most comfortable state for the electron corresponds to the case of $n = +1$, which is the minimum energy state (ground state). As n increases, $n = 2, 3, \ldots$, the electron finds itself in sequentially higher energy, excited states.

At this point I return to the electron in Fig. 1.1. The red light absorbed by the electron in liquid ammonia is associated with wavelengths λ longer than approximately 700 nm. On account of the duality, light can be viewed as both a wave and a particle. According to the particle theory, light is composed of discrete particles called photons whose energy is equal to $h\nu$. A photon with a wavelength of λ and a frequency ν proceeds in vacuum with a constant velocity of $c_0 = \lambda\nu \cong 3.0 \times 10^8\,\mathrm{m\,s^{-1}}$. Thus, the energy of a photon of $\lambda \cong 700\,\mathrm{nm}$ is calculated approximately as $h\nu = hc_0/\lambda \cong 0.28 \times 10^{-18}\,\mathrm{J}$ with the use of the Planck constant of $h \cong 6.6 \times 10^{-34}\,\mathrm{J\,s}$. Let us presume that the electron in liquid ammonia is excited by absorbing this amount of energy from the the ground state of $n = 1$ to the first excited state of $n = 2$. Then, substitution of $h\nu \cong 0.28 \times 10^{-18}\,\mathrm{J}$ and $m_\mathrm{e} \cong 9.1 \times 10^{-31}\,\mathrm{kg}$ into (1.6) allows an estimation of the wall-to-wall distance of the potential $2d$.

$$h\nu = \Delta E = E_2 - E_1 = (2^2 - 1^2)\frac{h^2}{32 m_\mathrm{e} d^2} \tag{1.6}$$

The distance is estimated as $2d \approx 0.8\,\mathrm{nm}$. To see whether this distance is reasonable or not, we assume the ammonia molecules are at each corner of a cubic lattice, with lattice constant l. Through knowledge of the density of liquid ammonia (approximately $0.7\,\mathrm{kg\,m^{-3}}$) and the mass of an ammonia molecule (approximately $2.8 \times 10^{-26}\,\mathrm{kg\,molecule^{-1}}$), the constant is estimated as to be $l \approx 3\,\mathrm{nm}$. Although the constant $l \approx 3\,\mathrm{nm}$ is about four times as large as the distance $2d \approx 0.8\,\mathrm{nm}$ above, it is of the same order of magnitude, crudely speaking. Thus, as a result, it can be proposed that the electron ejected from the alkali metal is in a cavity roughly the size of one ammonia molecule. Since such a situation is not unrealistic, it may be concluded that the phenomenon of the coloration is explicable in terms of the simplified square well potential, in place of the more complicated true potential.

The above example supports the statement of the title of Sect. 1.1. In other words, the energy state of an electron of a system can be probed by observing the absorption spectra of the system. Admittedly, the approximation of the square well potential in the above case is very crude. However, the Coulombic potential in the one-electron atomic system in Chap. 3 is real, not approximate, which, fortunately, allows the derivation of the exact energy states of the electron.

Incidentally, the above-mentioned conclusion that the electron in a square well potential cannot assume the state of $n = 0$ has a profound physical meaning in that it is a confirmation of the uncertainty principle introduced by W.K. Heisenberg. This principle demarcates intrinsically quantum mechanics and classical mechanics. In the former, a pair of physical quantities, e.g., position x and momentum p, cannot both be determined exactly at the same time, in contrast to classical mechanics. In the above instance, both x and p are associated with an intrinsic uncertainty Δx and Δp, and their values cannot vanish however precisely the measurement of both quantities is attempted. All that is ensured is the relation $\Delta x \Delta p \geq \hbar/2$.

I am a bit sorry to embarrass you by abruptly exposing one of the very basic principles of quantum mechanics, but for the moment, I ask you to accept that in quantum mechanics such a situation as simultaneous face-saving of two parties is impossible. To demonstrate the principle I will show several related examples in Sects. 2.1.12, 3.3.3, 3.4.1, 6.2, and 8.2.4. In the case of the square well potential, I will provide further explanation of the principle below.

As stated before, the energy E in the state with $n = 0$ becomes 0, so that according to the relation $E = p^2/(2m_e)$, p must be exactly 0, meaning $\Delta p = 0$. Therefore, the state is not allowed by the requirement of $\Delta x \Delta p \geq \hbar/2$. Actually, the energy of $E = p^2/(2m_e)$ is determined exactly by (1.5) not only for the case of $n = 0$, but also for cases of $n \neq 0$. However, in the latter cases, indetermination remains as to the *direction* of the vectorial quantity of momentum. What this means is, as explained in Sect. 2.1.12, that the sine function in (1.4) is an expression of the stationary standing wave obtained by the superposition of the right-going and the left-going waves, but in a standing wave expression, one cannot know the direction of the travelling waves. Similar situations will be encountered in Sects. 2.1.12, 3.3.3 and so on. In the first allowed state, $n = 1$, the energy is increased by E_1 from 0. This energy is called the zero-point energy.

The result in (1.4) stems from the forced boundary condition that the eigenfunction must be 0 at the point of $x = +d$. Without imposing this restriction, the number n need not be an integer, which leads to an infinite number of solutions. In such non-integer cases, e.g., $n = 1.5$ and 2.8 shown in Fig. 1.2(d), the eigenfunction $\phi(x)$ does not converge to 0 when the variable x approaches $+d$ from the left-hand side. However, the function $\phi(x)$ must be 0 in the regions of $|x| > d$ because the electron cannot exist there. As a result, both the function $\phi(x)$ and its first derivative with respect to x, $\mathrm{d}\phi(x)/\mathrm{d}x$, become discontinuous at x equal to $+d$. In quantum mechanics such a nonphysical solution is disregarded, that is to say, it is required that both $\phi(x)$ and $\mathrm{d}\phi(x)/\mathrm{d}x$ must be single-valued, finite, and continuous. It will be repeatedly shown that the physically acceptable energy states become discrete by imposing such boundary conditions (cf. Sects. 3.1, 3.2.2, 3.2.4 and so on).

As a continuation of this brief sketch of the electron, I will mention a little about the electron spin in Sect. 1.3, and then describe in Sect. 1.4 a short history of the birth of such equations as (1.1), which describe the energy state of the electron quantum mechanically.

Problems

1.1 We have learned that the ground state electron in liquid ammonia is described approximately by a function of $\sin\{\pi(x+d)/(2d)\}$, which becomes maximum at $x = 0$ and assumes the value of 0 at $x = \pm d$. Functions such as $(d^2 - x^2)$ and $e^{-x^2/d^2} - e^{-1})$ behave similarly. But they do not satisfy (1.4). Confirm this statement.

1.2 Derive the energy state function $\phi(x, y, z)$ and the associated energy E for an electron confined in a potential $V = 0$ for $|x| < d$, $|y| < d$, and $|z| < d$ and infinity for $|x| > d$, $|y| > d$, and $|z| > d$. Sketch the functions squared for the ground and the first excited states. Show that three different eigenfunctions are associated with the first excited state. (Hint: $\phi(x, y, z) \propto \phi(x)\phi(y)\phi(z)$, and the total energy is a simple sum of the three one-dimensional energies E_x, E_y, and E_z).

1.3 The Electron Spin: As Important as the Energy

In this section I will briefly discuss the electron focusing on the electron spin, a quantity which is as important as the energy described in Sects. 1.1 and 1.2. As a preliminary, I will start with the orbital angular momentum of the electron. Both the electron spin and the orbital angular momentum will be explored in detail in Chap. 3.

Any particle in a spherical potential like the electron in a hydrogen atom to be discussed in Chap. 3 is said to be in an isotropic field. For such a particle, the orbital angular momentum of the particle is a physical quantity as important as the energy. This is true in both classical and quantum mechanics. I will start with the familiar case of classical mechanics.

The orbital angular momentum $\boldsymbol{l}$ of a particle with mass m and velocity $\boldsymbol{v}$ is given by the cross product of the position vector $\boldsymbol{r} = (x, y, z)$ and the momentum vector $\boldsymbol{p} = m\boldsymbol{v} = (p_x, p_y, p_z)$ as illustrated in Fig. 1.3.

The cross product (outer product) of vectors is defined as a vector whose absolute value is equal to the value of area of a parallelogram determined by the two vectors $\boldsymbol{r}$ and $\boldsymbol{p}$ and whose positive direction is taken as equal to the direction that a right-handed screw progresses when the vector $\boldsymbol{r}$ is turned toward $\boldsymbol{p}$ via the smaller of the two angles between them.

Thus, if $\boldsymbol{p}$ is decomposed into two components, parallel $\boldsymbol{p}_\parallel$ and perpendicular $\boldsymbol{p}_\perp$ as Fig. 1.3 shows, the magnitude of $\boldsymbol{l}$ becomes equal to $rp_\perp$ with $p_\perp \equiv |\boldsymbol{p}_\perp|$ and the direction of $\boldsymbol{l}$ is upward.

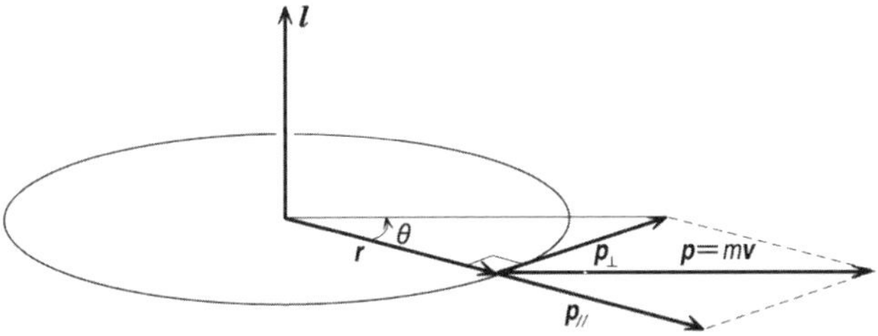

Fig. 1.3. Orbital angular momentum of a particle

More generally, it can be shown by geometrical construction that $l_x = yp_z - zp_y$, $l_y = zp_x - xp_z$, $l_z = xp_y - yp_x$, where $\boldsymbol{l} = (l_x, l_y, l_z)$. A mnemonic technique to remember these relations can be constructed from the fact that $\boldsymbol{l}$ can be expressed by the following determinant where $\boldsymbol{i}, \boldsymbol{j}, \boldsymbol{k}$ stand for the unit vectors of Cartesian coordinates.

$$\begin{vmatrix} \boldsymbol{i} & \boldsymbol{j} & \boldsymbol{k} \\ x & y & z \\ p_x & p_y & p_z \end{vmatrix}$$

In Fig. 1.3, the vector $\boldsymbol{l}$ is expressed as $(0, 0, l_z)$ by choosing the plane $\boldsymbol{r} - \boldsymbol{p}$ as the xy plane, and l_z is given as $rp_\perp$. This product is allowed to take continuous values in classical mechanics, of course.

If the particle in Fig. 1.3 possesses an electrical charge in addition to the mass, the particle will then be associated with a magnetic moment $\boldsymbol{\mu}$ proportional to $\boldsymbol{l}$, i.e., $\boldsymbol{\mu} = \gamma \boldsymbol{l}$ in accordance with classical electromagnetism. Here, the proportionality constant γ is called the magnetogyric ratio, the dimension of which is equal to (quantity of electricity)/(mass). The magnetic moment can be regarded as a tiny magnet so that in a field of magnetic flux density $\boldsymbol{B}$, the moment gains a potential energy of $E = -\boldsymbol{\mu} \cdot \boldsymbol{B}$. If the direction of the field is taken to be in the z-axis, the magnetic flux density is written as $\boldsymbol{B} = (0, 0, B)$. Thus, the energy becomes as $E = -\mu_z B = -\gamma l_z B$ where μ_z is the z-component of the magnetic moment. This relation implies that the 'magnet' can take any value of the potential energy according to the value of l_z.

Now, at this point I will replace the above classical particle with an electron. In this case also, the electron acquires an orbital angular momentum $\boldsymbol{l}$ and a magnetic moment $\boldsymbol{\mu}$ as a consequence of its spatial motion. However, as will be shown in Chap. 3, the values for $\boldsymbol{l}$ and its component l_z cannot be allowed to vary continuously as in the case of the classical particle discussed above. This situation is reminiscent of the fact that the energy of an electron cannot change continuously as described in Sect. 1.2. If the conclusion in Sect. 3.2.3 is taken in advance, the allowed value of $\boldsymbol{l}$ must be restricted to any one of those given by $\sqrt{l(l+1)}\hbar$ with l being $0, 1, 2, 3, \ldots$. Also, the

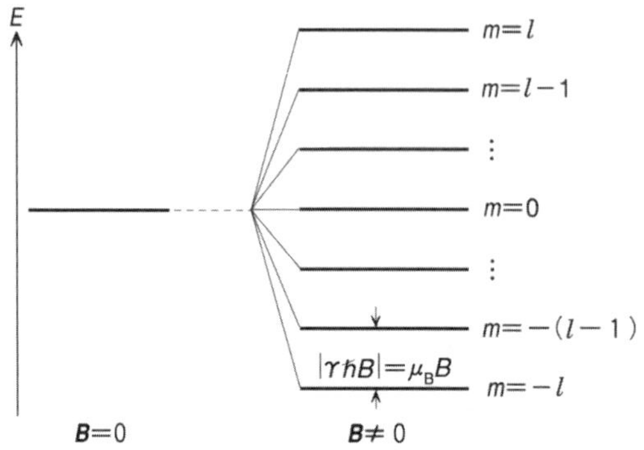

Fig. 1.4. Energy splitting due to interaction of the magnetic moment originating from orbital motion of the electron with the magnetic flux density

value of l_z is limited to $m\hbar$ where m runs over $0, \pm 1, \pm 2, \pm 3, \ldots, \pm l$ (note that the symbol m here has nothing to do with mass).

According to the results so far stated, the energy of an electron in a spherically symmetric Coulombic potential is split into $2l + 1$ different states with energies of $E = -\gamma l_z B = -\gamma m \hbar B$ when a magnetic field is applied in addition to the spherical electrostatic field.

Figure 1.4 demonstrates such a splitting, where μ_B is named the Bohr magneton, which is the unit of magnetic moment expressed as $\mu_\mathrm{B} = e\hbar/(2m_\mathrm{e})$.

The constant γ is a negative quantity which originates from the fact that the electron is negatively charged. Therefore, the energy E becomes a minimum for $m = -l$. The splitting depicted in Fig. 1.4 illustrates the normal Zeeman effect, which has been known empirically since roughly the end of the nineteenth century through observation of splittings in atomic line spectra and other phenomena.

Later on, however, it was revealed that the electron possesses another type of angular momentum, which has nothing to do with spatial motion, and thus, no analogue in classical mechanics: in 1925, G.E. Uhlenbeck and S.A. Goudsmit discovered that the atomic lines of a gaseous alkali metal spectrum split into two components, which is inexplicable if one considers only the normal Zeeman effect.

In order to account for this startling finding they introduced a new concept which stated that the electron contains intrinsically a new type of angular momentum having no relation to its spatial motion. With the introduction of this new concept, previously known but unexplained strange experimental results were consistently accounted for. Among such 'strange' results one may consider the so-called anomalous Zeeman effect which, along with the normal Zeeman effect, are credited to P. Zeeman in the late nineteenth century. This anomaly meant 'extra' splittings of atomic line spectra in a magnetic field which were unexplained in terms of the normal Zeeman effect (cf. related topic in Sect. 4.3.1). Another example is the seemingly odd splitting of the atomic silver beam in the magnetic field into two components, which was first discovered by O. Stern and W. Gerlach in 1922 (cf. the electronic shell structure of the ground-state silver atom in Table 4.1). All these ‘odd’ phenomena were explained consistently through the introduction of the electron spin.

The newly introduced angular momentum fits in perfectly with quantum mechanics. In order to be consistent with experiment results, the value of this new angular momentum must be restricted to only $+\hbar/2$ and $-\hbar/2$. This angular momentum is sometimes called the ‘intrinsic angular momentum’ of the electron in the sense that it is inherently associated with the electron, but more generally, it is called the electron spin as introduced at the beginning of this section. The two states, $+\hbar/2$ and $-\hbar/2$, are indistinguishable in the absence of a magnetic field. Only in the field can they be differentiated, just like a tiny magnet which is either parallel or anti-parallel relative to the field.

At this point, I will return to the system of an electron in liquid ammonia shown in Fig. 1.1. The ground-state electron in liquid ammonia is stabilized with no orbital angular momentum. Since the magnetic moment originating from the spatial motion of the electron is 0, there may seem to be no splitting due to the normal Zeeman effect as depicted in Fig. 1.4, not to mention any transition among the energy states split by an externally applied magnetic field. Actually, however, the system does show a transition in the presence of a magnetic field, which indicates that the energy states are split and the electron is associated with a magnetic moment even though the orbital angular momentum is absent. This magnetic moment is due to nothing else but the intrinsic angular momentum of electron, i.e., the electron spin. To be more specific, when the system is placed in a magnetic field parallel to the z-axis with a magnetic flux density of, e.g., approximately $0.3\,\mathrm{T}$ (tesla) ($\mathrm{T} = \mathrm{V\,s\,m^{-2}}$), and is irradiated with an electromagnetic wave with frequency ν approximately equal to $8.4 \times 10^9\,\mathrm{Hz}$ (microwave), an absorption of part of the magnetic component of the microwave is observed.

The energy of the absorbed microwave is calculated as $h\nu = 5.5 \times 10^{-24}\,\mathrm{J}$, with the above frequency and the Planck constant of $h \approx 6.6 \times 10^{-34}\,\mathrm{J\,s}$. The magnitude of this energy is 5 to 6 orders of magnitude smaller than the energy of the red light absorbed by the electron, which was calculated crudely as $h\nu \approx 10^{-18}\,\mathrm{J}$ (cf. Sect. 1.2). Therefore, the energy absorbed by the electron spin in the magnetic field above may appear negligibly small. However, as will be shown in Sect. 4.1, the electron spin plays a fundamental role, via the Pauli exclusion principle, in various aspects of chemistry. In particular, in the construction of the electronic shell structure of multi-electron atoms (Chap. 4) and chemical bond formation (Chap. 7 and on). The electron spin is truly as important a thing as the energy in governing the electron's behavior.

However, since the electron spin is a very quantum mechanical phenomenon, I will try to explain it in various ways throughout the book to assist the reader in understanding its nuances.

1.4 A Brief History of the Quantum Mechanical Description of the Electron

In this section I will explain, in two parts, how the equation to describe the electron was derived, first, by introducing the Bohr theory, and then, the Schrödinger theory. Contrary to mathematics and philosophy, chemistry as well as physics are to be constructed on the basis of experiment and empiricism. The Schrödinger equation, which is the very basis of modern chemistry, was derived through the outcome of a chain of experiments, hypotheses, and logics. Although the purpose of this book is not to cover the history of quantum mechanics, a brief look at the trial and error process in the derivation of the Schrödinger equation should give the reader a better feel for the theory.

1.4.1 The Birth of the Bohr Theory

Ever since J. Dalton postulated his atomic theory early in the nineteenth century, people steadily quantified the characteristics of macroscopic substances (elements) by precise measurement of mass and gas volume to determine the *relative masses* of elements involved in chemical events, which culminated in the determination of atomic *ratios* in molecules. As a consequence of this effort, chemistry as a primary material science grew up, so much so that by the end of the nineteenth century, its level was comparable to that of modern chemistry to a considerable extent, in so far as how matter was treated as a macroscopic subject.

However, the view that atoms are microscopic particles was meticulously set aside by F.W. Ostwald in his influential textbook (*Grundlinien der anorganische Chemie*, 1900). Likewise, E. Mach, an eminent physicist and philosopher, was also critical of the view that matter (elements) consists of discrete particles.

On the other hand, R.W. Bunsen and G.R. Kirchhoff analyzed the flames of substances with the use of a prism late in the nineteenth century. They discovered that each element exhibits its own characteristic spectra at certain visible wavelengths. The discovery led to the establishment of the spectroscopic technique for the identification of elements. Simultaneously, much effort was made to determine accurately the absolute wavelengths of the light spectra, which was essential for the spectroscopic method.

In parallel with the spectroscopic analyses undertaken in laboratories, astronomers also performed spectroscopic measurements on celestial phenomena and found that the results of both the laboratory and celestial measurements coincided nicely. For instance, J.J. Balmer found in 1885 that the wavelength of some lines from a celestial observation agreed with the line spectra observed for discharged hydrogen gas in the laboratory. He noticed that the wavelength λ of coinciding lines obeyed the following relationship:

$$\frac{1}{\lambda} = \frac{1}{91.14\,\mathrm{nm}} \left(\frac{1}{2^2} - \frac{1}{n^2} \right) \quad (n \geq 3) \tag{1.7}$$

This relation was generalized by J. Rydberg in 1890 in the form shown as (1.8) below, which was found to be applicable to the line spectrum of not only hydrogen, but also many other elements, in particular, alkali metals.

$$\frac{1}{\lambda} = R_\infty \left\{ \frac{1}{(m+\mu)^2} - \frac{1}{(n+\nu)^2} \right\} \tag{1.8}$$

In (1.8), the letters m and n are positive integers ($m < n$), μ and ν are constants proper to each element, and the symbol R_∞ is identified as the Rydberg constant at the present time, the value of which is close to the reciprocal of 91.14 nm. Relations such as shown in (1.7) and (1.8) were known to people even before the recognition that elements consist of microscopic particles. The relationships remained merely empirical until the real physical consequences were discovered, thanks to Bohr as will be discussed shortly.

A turning point was the discovery by E. Rutherford in 1911 who demonstrated that alpha particles (soon to be recognized as He^+ ions) impinging on thin films of gold were occasionally scattered with an unexpectedly large angle. From this result, Rutherford imagined correctly that an atom is a microscopic particle consisting of a tiny, massive, and positively charged nucleus and negatively charged electrons that are circulating the nucleus. The electron had been detected by J.J. Thomson a little earlier (1897) than the experiments of Rutherford. A model somewhat similar to Rutherford's had previously been proposed by H. Nagaoka (1904), but regrettably, the model was based on totally classical theories.

N.H.D. Bohr, learning of Rutherford's experiments, as well as the empirical relations known in the circle of astronomy, took the problem of the atom seriously. He regarded the hydrogen atom as being composed of a proton and an orbiting electron, which had a similarity to the Nagaoka model. Contrary to those who remained mired in classical theories, Bohr attempted to leap beyond. First, he assumed a balance of forces acting on the electron in accordance with classical electrodynamics and classical mechanics. The balance was expressed in terms of the Coulombic attraction (the left-hand side of (1.9)) and the centrifugal force (the right-hand side of (1.9)) as

$$\frac{e^2}{4\pi\varepsilon_0 r^2} = \frac{m_e v^2}{r} \tag{1.9}$$

where the symbols m_e, e, v, and r stand for the mass, electric charge, velocity, and orbital radius of the electron, respectively, ε_0 is the dielectric constant of vacuum.

Second, Bohr adopted the hypothesis of 'quantization' advocated by Planck in 1900. Bohr expressed this quantization as:

$$\int_0^{2\pi} r_n m_e v \mathrm{d}\theta = 2\pi r_n m_e v = nh \quad \text{where } n = 1, 2, 3, \ldots \tag{1.10}$$

That is to say, Bohr assumed that the electron could stay stable only when the radius of the orbit satisfied the above discrete condition. Since the middle of the above equation is equal to an orbital angular momentum multiplied by 2π, the equation requires that the orbital angular momentum of the electron must be multiples of $h/2\pi$ (cf. Fig. 1.3). This is indeed 'quantization' of the orbital angular momentum, which could change continuously in classical mechanics. From (1.9) and (1.10) the radius of orbit r_n and the velocity v are derived, and substitution of these results into the expression for the energy, E_n = (kinetic energy) + (potential energy) $= m_e v^2/2 - e^2/(4\pi\varepsilon_0 r_n)$, leads to the following result, which is also discrete.

$$E_n = -\frac{m_e e^4}{2(4\pi\varepsilon_0\hbar)^2} \times \frac{1}{n^2} \tag{1.11}$$

Third, Bohr postulated that a transition between energy states was possible only when the energy difference between the states matched the energy of a photon $h\nu$ in such a way as to satisfy the equation, $\Delta E = E_n - E_m = h\nu$. On the basis of these assumptions Bohr was successful in explaining the observed line spectra of the hydrogen atom in 1913.

The light corresponding to an atomic hydrogen line with a wavelength of λ and a frequency ν is associated with the velocity c_0 in vacuum as $c_0 = \lambda\nu$ and has an energy of $h\nu = hc_0/\lambda$. Thus, from the relation, $\Delta E = E_n - E_m = h\nu = hc_0/\lambda$, the following equation is derived.

$$\frac{1}{\lambda} = R_\infty\left(\frac{1}{m^2} - \frac{1}{n^2}\right), \quad R_\infty = \frac{m_e e^4}{2hc_0(4\pi\varepsilon_0\hbar)^2} \tag{1.12}$$

This equation has the same form as the empirical equation in (1.7). In addition, the constant R_∞, which used to be a mere empirical constant, is now endowed with a physical meaning in terms of the five universal constants. As to the radius r_n derivable from (1.9) and (1.10), $r_1 = 4\pi\varepsilon_0\hbar^2/(m_e e^2)$ is calculated as $r_1 = 5.29 \times 10^{-11}$ m. This radius is called the Bohr radius and is denoted as a_0. As will be shown in Sect. 3.2.4, this radius as well as the energy E_1 obtained for $n = 1$ in (1.10) agree exactly with those calculated from the new theory introduced by Schrödinger after Bohr.

In passing, the velocity of the electron in state $n = 1$ is calculated as $v \approx 2.18 \times 10^6\,\mathrm{m\,s^{-1}}$, which will be discussed shortly. Bohr's success in explaining the spectrum of the atomic hydrogen was the dawn of new era, the era of quantum theory into which people advanced. However, Bohr's theory was still incomplete, and another decade or so was necessary before the advent of modern quantum theory, the birth of which is described in the next section (cf. Rigden in Refs. for Chap. 1).

1.4.2 The Birth of the Schrödinger Theory

Bohr's older quantum theory was soon replaced by a new quantum theory in the form of matrix mechanics due to W.K. Heisenberg, M. Born, and P. Jordan as well as in the form of wave mechanics due to L.-V.de Broglie and

E. Schrödinger. The two approaches are in fact equivalent despite their differing appearance, this fact having been pointed out by Schrödinger himself in 1926. At present, quantum mechanics is pervasive throughout chemistry. Most problems are solved by way of matrix mechanics because of its superior adaptability to computers. For ease of conceptually grasping the physical significance, however, wave mechanics has the advantage over matrix mechanics, with the proviso that the reader is prepared to accept mathematical manipulations of operators. In order to aid in understanding this issue, I will repeat the key points in this section and in Chaps. 2 and 3. The reader may also find some explanation on mathematical operators in Sects. 2.1.12, 3.2.2, and 8.1.1.

Before returning to the story of the birth of the Schrödinger equation, I should like to draw attention to the following point. As mentioned, the spectrum of atomic hydrogen led Bohr to develop his quantum theory. However, it should be pointed out that the experimental design necessary to perform accurate spectral observations of atomic hydrogen is a difficult task, resulting from the fact that when one uses the electric discharge of hydrogen molecules to generate atomic hydrogen, there is a great difficulty in differentiating the spectral signal due to the H atom from the signal due to the H_2 molecule. Since the latter is much more complicated than the former, correctly selecting the atomic hydrogen signal required experimental patience and craft. Only after such efforts, together with Bohr's genius, was the new horizon reached.

As mentioned in the end of the preceding section, the velocity of the electron at state $n = 1$ is calculated as $v \approx 2.18 \times 10^6\,\mathrm{m\,s^{-1}}$. This is much smaller than the velocity of light in vacuum, $c_0 \approx 3.0 \times 10^8\,\mathrm{m\,s^{-1}}$, but is not vanishingly small. In his attempt to advance Bohr's theory, Schrödinger at first stuck to this point and tried to develop a new theory compatible with the general principle of relativity of A. Einstein, which had already become a firm foundation of natural science around 1905. According to the principle of relativity, the following relation holds where the symbols m_e, p, and E stand for the mass of an electron at rest, the momentum and energy of an electron, respectively.

$$E^2 = p^2 c_0^2 + m_\mathrm{e}^2 c_0^4 \tag{1.13}$$

In addition to the principle of relativity Schrödinger was also concerned about the proposal of de Broglie put forth in 1924. L.-V. de Broglie intuitively considered that the electron, a corpuscular entity, might also behave as a wave. This is congruous with the dual particle and wave characterizations of light, greatly argued since the time of Newton. L.-V. de Broglie started with light having a frequency ν, an energy $E = h\nu$, and a momentum $p = E/c_0$, defined according to the requirements of the principle of relativity. Combining these relations leads to the following relationship between the momentum and the wavelength λ of light:

$$p = E/c_0 = h\nu/c_0 = h/\lambda \tag{1.14}$$

Since the momentum of an electron viewed as a particle is expressed as $\boldsymbol{p} = m\boldsymbol{v}$ in (1.14), de Broglie was inspired to imagine that an electron could be viewed as a wave, leading to the following relation, called de Broglie's equation.

$$\lambda = \frac{h}{|\boldsymbol{p}|} \tag{1.15}$$

Schrödinger adopted this view of de Broglie and supposed that matter, such as an electron, could be described in terms of a 'material wave', defined in terms of a wave function ψ, though Schrödinger did not have clear cognizance of the physical meaning of the function at the beginning.

In efforts to define this wave function, Schrödinger first resorted to the classical expression defining three-dimensional plane waves. The amplitude $\Psi(\boldsymbol{r}, t)$ of such a wave may be written, e.g., as $\Psi(\boldsymbol{r}, t) = \Psi_0 \cos 2\pi(\boldsymbol{k} \cdot \boldsymbol{r} - \nu t)$ with $\boldsymbol{r} = (x, y, z)$ to obtain the following equation:

$$\frac{\partial^2 \Psi}{\partial t^2} = (\lambda\nu)^2 \left(\frac{\partial^2}{\partial x^2} + \frac{\partial^2}{\partial y^2} + \frac{\partial^2}{\partial z^2} \right) \Psi \tag{1.16}$$

Here, λ is the wavelength of a magnitude $\lambda = 1/|\boldsymbol{k}| \equiv 1/k$, ν is the frequency per unit time, and the product $\lambda\nu$ is equal to the velocity of the wave. The vector $\boldsymbol{k}$ having the magnitude of $k = 1/\lambda$ is sometimes called the wave number vector.

Referring to the relation in (1.16), Schrödinger conceived that the wave function ψ supposed for an electron should obey a relation similar to (1.16) as much as possible. He conjectured that in the equation for ψ the wavelength λ appearing in (1.16) was to be replaced by the relation $\lambda = 1/k$ with k being given by de Broglie's equation in (1.15), i.e., $k = |\boldsymbol{p}|/h$. This equation links the two expressions of the momentum of an electron, $\boldsymbol{p}$ regarded as a particle and $h\boldsymbol{k}$ as a wave.

At this stage, Schrödinger took notice of the fact that the relation in (1.13) could be recovered if the energy E and the momentum $\boldsymbol{p}$ (consisting of the components, p_x, p_y, and p_z) represent the operational functions implied in (1.17) and the squared quantities E^2 and p^2 represent repetitive operations.

$$E \to \mathrm{i}\hbar\frac{\partial}{\partial t}, \quad \boldsymbol{p} \to p_x = \frac{\hbar}{\mathrm{i}}\frac{\partial}{\partial x}, \quad p_y = \frac{\hbar}{\mathrm{i}}\frac{\partial}{\partial y}, \quad p_z = \frac{\hbar}{\mathrm{i}}\frac{\partial}{\partial z} \tag{1.17}$$

The repetitive operations turn out to be such that $E^2 \to (\mathrm{i}\hbar\partial/\partial t)(\mathrm{i}\hbar\partial/\partial t) = -\hbar^2\partial^2/\partial t^2$, for example. Thus, in connection with the wave function ψ, (1.13) is rewritten as follows:

$$\frac{\partial^2 \psi}{\partial t^2} = c_0^2 \left(\frac{\partial^2}{\partial x^2} + \frac{\partial^2}{\partial y^2} + \frac{\partial^2}{\partial z^2} \right) \psi - \frac{m_\mathrm{e}^2 c_0^4}{\hbar^2}\psi \tag{1.18}$$

It can be seen that the first term of the right-hand side of (1.18) has essentially the same form as the right-hand side of (1.16) for a classical wave. Since the second term of the right-hand side of (1.18) is a mere multiplication of ψ with universal constants, it seemed that (1.18) was an appropriate equation to determine the sought-after wave function. Schrödinger reached this point, but he refrained from continuing along this line because of trouble with the interpretation of the wave function ψ and because of difficulties in knowing how to include the electromagnetic potential.

Schrödinger next retreated to find an equation valid in the non-relativistic limit. He restarted from the limit of $pc_0 \ll m_e c_0^2$ in (1.13) and obtained the following equation:

$$E = \sqrt{p^2 c_0^2 + m_e^2 c_0^4} \cong m_e c_0^2 + \frac{p^2 c_0^2}{2 m_e c_0^2} = m_e c_0^2 + \frac{p^2}{2 m_e} \quad (1.19)$$

If energy is measured from 0 taken at the value $m_e c_0^2$ and if the case $V \neq 0$ is also taken into account, the energy E can be expressed as follows:

$$E = \frac{p^2}{2 m_e} + V \quad (1.20)$$

At this stage, if E and p are interpreted again as representing the operational functions in (1.17), the following equation is obtained which should determine the wave function ψ for the electron.

$$\mathrm{i}\hbar \frac{\partial \psi}{\partial t} = -\frac{\hbar^2}{2 m_e} \left(\frac{\partial^2}{\partial x^2} + \frac{\partial^2}{\partial y^2} + \frac{\partial^2}{\partial z^2} \right) \psi + V\psi \quad (1.21)$$

The sum in parenthesis above is an operational function called the Laplacian, which has long been known in mathematics and physics. The function is customarily denoted as follows:

$$\left(\frac{\partial^2}{\partial x^2} + \frac{\partial^2}{\partial y^2} + \frac{\partial^2}{\partial z^2} \right) \psi = \Delta \psi \quad (1.22)$$

The symbol of the Laplacian Δ may be written as $\Delta = \nabla\nabla = \nabla^2$ (with the use of another operator called nabla ∇). This operator is equal to the sum of $\boldsymbol{i}\partial/\partial x + \boldsymbol{j}\partial/\partial y + \boldsymbol{k}\partial/\partial z$, where $\boldsymbol{i}$, $\boldsymbol{j}$ and $\boldsymbol{k}$ stand for the unit vectors along the x, y, and z axes, respectively.

Problem

1.3 Verify the relation $\Delta = \nabla\nabla = \nabla^2$.

For simplicity, the right-hand side of (1.21) is written as (1.23)

$$\mathrm{i}\hbar\frac{\partial\psi}{\partial t} = \hat{H}\psi \tag{1.23}$$

$$\hat{H} \equiv -\frac{\hbar^2}{2m_\mathrm{e}}\left(\frac{\partial^2}{\partial x^2}+\frac{\partial^2}{\partial y^2}+\frac{\partial^2}{\partial z^2}\right)+V \tag{1.24}$$

On account of the derivation procedure, the above equation (1.23) is called the wave equation. Alternatively, it is also called the time-dependent Schrödinger equation. It was proposed in 1926 by Schrödinger. As to the physical meaning of the wave function ψ, which was unclear to Schrödinger himself at the beginning, Born among others, soon interpreted correctly that the square of the absolute value of the function is proportional to the probability of the presence of an electron. This interpretation was confirmed by comparison with the results of an experiment of an electron scattering by atoms, and by many other pieces of supporting evidence. The function is popularly called the wave function because it is a solution of the wave equation.

Presently, the Schrödinger equation (1.23) is the most fundamental description of the electron. Yet, as shown above, it is the outcome of trial and error and it cannot be verified like mathematical theorems. Its justification depends totally upon the ability to prove and predict experimental events. Such artificial substitutions of physical quantities with operational functions as in (1.17) are also justified only by the fact that the substitutions lead to consistency with experiment.

It is noticed at this point that the wave function derived above involves only the spatial coordinates and time as variables, giving no information as to the electron spin mentioned in Sect. 1.3. As stated in that section, the electron spin can assume only the values of $+1/2$ or $-1/2$ in units of $\hbar$. This is a demand totally dictated by experimental results. In order to include the electron spin so as to be compatible with (1.23), W. Pauli put forth an interim theory in 1927. According to that theory the solution of (1.23) is interpreted as expressing equally both states of spin, i.e., the two states $\psi(x, y, z, t, +1/2)$ and $\psi(x, y, z, t, -1/2)$ satisfy equally the equation with regard to the spatial coordinate and time (cf. Sect. 3.4.3).

The Schrödinger-Pauli theory is a makeshift to accommodate the electron spin at the non-relativistic limit approached by Schrödinger. The first theory compatible with the principles of relativity is the accomplishment of P.A.M. Dirac in 1928. If spectroscopic resolution is greatly enhanced, very fine spectral structures originating from the electron spin and other relativistic effects are revealed even in the lightest H atom. In order to explain the discrepancy between spectroscopic measurements and the simple predictions by (1.7) one has to employ the theory by Dirac and/or modern quantum electrodynamics theory.

Fortunately, however, most problems in chemistry related to electron spin can be dealt with by application of the Schrödinger-Pauli theory. In cases where the relativistic effect is not negligible, theoretical techniques based on perturbation theory (cf. Sect. 2.1), or the effective potential approximation*, have been developed. Chemistry would have become much more difficult, if in all chemical problems one had only the highly sophisticated theories like Dirac's.

Problem

1.4 Confirm that the function $\Psi(\boldsymbol{r}, t) = \Psi_0 \cos 2\pi(\boldsymbol{k} \cdot \boldsymbol{r} - \nu t)$ is a solution of (1.16). By referring to the definition of partial differentiation show that any function of the form $f(ax + bt)$ satisfies the wave equation in (1.16). Here, the symbols x, t, and a, b represent distance, time, and arbitrary constants, respectively.

2 Fundamentals of Quantum Mechanics: A Prerequisite for Understanding the Chemical Bond

I have already discussed without explanation such quantum mechanical terms as the Schrödinger equation, the Hamiltonian operator, boundary conditions, the uncertainty principle, and so on. To make up for this cursory introduction in Chap. 1, I will present a summary of explanations in this chapter on the basic concepts, postulates, definitions, and methodology proper to quantum mechanics. The first half of the chapter (Sect. 2.1) is devoted to general explanations of these fundamental subjects. In the latter half (Sect. 2.2) some practical examples of the fundamentals are illustrated in the case of the square well model potential.

2.1 Basic Concepts, Postulates, Definitions and Methodology
Fundamental concepts will be explained which include the meaning of the wave function, eigenstates, superposition of states, the definition and properties of Hermitian operators, and approximations by both the perturbation theory and by the calculus of variation.

2.2 Application of the Concepts Presented in Sect. 2.1 to the Problem of the Square Well Potential

2.1 Basic Concepts, Postulates, Definitions and Methodology

2.1.1 Wavefunction Determines the Probability of Finding the Electron

In classical Newtonian mechanics the motion of a macroscopic particle at an arbitrary time can be predicted precisely from the knowledge of coordinates $\boldsymbol{r}$ and momentum $\boldsymbol{p}$ at a certain instance. In contrast, in quantum mechanics dealing with microscopic particles like the electron, the description of the state of the particles starts from the time dependent-Schrödinger equation (1.23) with the definition of (1.24) in Chap. 1, which are reproduced below.

$$\mathrm{i}\hbar\frac{\partial\psi}{\partial t} = \hat{H}\psi \tag{1.23}$$

$$\hat{H} \equiv -\frac{\hbar^2}{2m_\mathrm{e}}\left(\frac{\partial^2}{\partial x^2} + \frac{\partial^2}{\partial y^2} + \frac{\partial^2}{\partial z^2}\right) + V \tag{1.24}$$

Shortly after the discovery of (1.23) by Schrödinger, Born postulated that the probability of finding the particle in the region between $\boldsymbol{r}$ and $\boldsymbol{r}+\mathrm{d}\boldsymbol{r}$ at time t is given by the product of the square of the absolute value of $\psi(\boldsymbol{r},t)$, denoted by $|\psi(\boldsymbol{r},t)|^2$, and the volume element $\mathrm{d}v(\mathrm{d}v = \mathrm{d}x\mathrm{d}y\mathrm{d}z)$. The square of the absolute value of $\psi(\boldsymbol{r},t)$ is named the probability density. As is foreseeable from (1.23), the function $\psi(\boldsymbol{r},t)$ can be a complex. The square of the absolute value of a complex quantity is given by the product of that quantity with its complex conjugate, which is formed by replacing $\mathrm{i} = \sqrt{-1}$ by $-\mathrm{i}$ wherever it occurs. This product always becomes a real number. When the real number $\psi^*(\boldsymbol{r},t)\psi(\boldsymbol{r},t) \doteq |\psi(\boldsymbol{r},t)|^2$ is integrated over all space, the integral $\int_{-\infty}^{\infty}\psi^*(\boldsymbol{r},t)\psi(\boldsymbol{r},t)\mathrm{d}v$ must be unity because of the meaning of probability. Therefore, the dimension of the wave function $\psi(\boldsymbol{r},t)$ is equal to the reciprocal of the square root of volume. Equation (1.23) is interpreted to give the time dependence of the wave function at an arbitrary time.

By now it should be clear for the reader that the dimension of the amplitude shown in Fig. 1.2(c) is equal to the reciprocal of the square root of length because the wave function in that figure is for a one-dimensional square well potential.

Dirac regarded the two functions $\psi(\boldsymbol{r},t)$ and $\psi^*(\boldsymbol{r},t)$ as two vectors in an abstract functional space and called the respective functions the ket vector and the bra vector. He proposed the notations $|\psi\rangle$ and $\langle\psi|$ for these vectors with which the definite integral $\int_{-\infty}^{\infty}\psi^*(\boldsymbol{r},t)\psi(\boldsymbol{r},t)\mathrm{d}\nu$ is written concisely as follows.

$$\int_{-\infty}^{\infty}\psi^*(\boldsymbol{r},t)\psi(\boldsymbol{r},t)\mathrm{d}v \equiv \langle\psi|\psi\rangle = 1 \tag{2.1}$$

2.1.2 Energy Eigenstates are Stationary

When the potential V in (1.24) is independent of time, i.e., $V = V(\boldsymbol{r})$, it is guaranteed that the energy E remains constant. In such a case the wave function $\psi(\boldsymbol{r},t)$ of a system is known to be a simple product of a spatial component $\phi(\boldsymbol{r})$ and a time dependent component $T(t)$. The spatial component is determined by the following eigenvalue equation, which is called the time-independent Schrödinger equation:

$$\hat{H}\phi(\boldsymbol{r}) = E\phi(\boldsymbol{r}) \tag{2.2}$$

The above equation is derived as follows: substitution of $\psi(\boldsymbol{r},t) = \phi(\boldsymbol{r})T(t)$ into (1.23) reproduced above gives the following relation,

$$\mathrm{i}\hbar\frac{\partial\psi(\boldsymbol{r},t)}{\partial t} = \mathrm{i}\hbar\phi(\boldsymbol{r})\frac{\mathrm{d}T(t)}{\mathrm{d}t} = \hat{H}\phi(\boldsymbol{r})T(t) = \left[-\frac{\hbar^2}{2m_\mathrm{e}}\Delta + V\right]\phi(\boldsymbol{r})T(t)\,. \tag{2.3}$$

Comparison of the second and the third terms leads to the relation below.

$$\mathrm{i}\hbar\frac{\left(\dfrac{\mathrm{d}T(t)}{\mathrm{d}t}\right)}{T(t)}=\frac{\hat{H}\phi(\boldsymbol{r})}{\phi(\boldsymbol{r})} \tag{2.4}$$

The left-hand side is a function of t alone whereas the right-hand side depends only on the spatial coordinate $\boldsymbol{r}$. Thus, in order for both sides to be always equal they have to be a constant independent of both variables t and $\boldsymbol{r}$. It is obvious from the left-hand side that the dimension of this constant is equal to the dimension of energy. Denoting this constant by E, one can derive the following two equations from (2.4), one of which is (2.2) itself.

$$\mathrm{i}\hbar\frac{\mathrm{d}T(t)}{\mathrm{d}t}=ET(t) \tag{2.5}$$

$$\hat{H}\phi(\boldsymbol{r})=E\phi(\boldsymbol{r}) \quad \text{with the identity,} \quad \hat{H}\equiv-\frac{\hbar^2}{2m_\mathrm{e}}\Delta+V \tag{2.6}$$

The above procedure to decompose a single equation (2.3) into two equations (2.5) and (2.6) for the variables t and $\boldsymbol{r}$ is commonly known as the method of separation of variables. Equation (2.5) is immediately solved as $T(t)=\exp(-\mathrm{i}Et/\hbar)$. As for (2.6), it is an eigenvalue equation introduced in Sect. 1.2 because it requires that the result of the operation of $\hat{H}$ upon the function $\phi(\boldsymbol{r})$ must be equal to a simple multiple of the same function. This eigenvalue equation for energy is called the time-independent Schrödinger equation, as stated above.

From the solutions of (2.5) and (2.6), the wave function $\psi(\boldsymbol{r},t)=\psi(\boldsymbol{r})T(t)$ is written as $\psi(\boldsymbol{r},t)=\exp(-\mathrm{i}Et/\hbar)\phi(\boldsymbol{r})$, from which $|\psi(\boldsymbol{r},t)|^2$ becomes equal to the time-independent $|\phi(\boldsymbol{r})|^2$. Because of this time-independence, the energy eigenstate is called a stationary state. Notice that this conclusion originates from the aforementioned condition that the potential V of the system under consideration is time-independent. When the system is in an energy state above the ground state of minimum energy, however, the system is known to emit spontaneously electromagnetic radiation with a lifetime characteristic of that state. Therefore, in these states the system remains stationary only in so far as the emission of radiation is negligible.

The solution of (2.6), $\phi(\boldsymbol{r})$, does not depend upon time contrary to the wave function of $\psi(\boldsymbol{r},t)$ in (1.23). Therefore, the function $\phi(\boldsymbol{r})$ is properly called the 'energy eigenfunction' or the 'energy state function'. However, the function $\phi(\boldsymbol{r})$ is often called the wave function. This is because the function has a wave-like form with regard to the spatial coordinate as shown, e.g., in the case of the square well potential (cf. Sect. 1.1). It is so called also because the function is derived from (1.23), which originated from the material wave theory for the electron as described in Sect. 1.4.2. In this book I will call functions depending on both $\boldsymbol{r}$ and t wave functions, while those depending only on $\boldsymbol{r}$ will be called energy eigenfunctions.

2.1.3 Any Physical Quantity has a Corresponding Operator which has its own Set of Eigenfunctions and Eigenvalues

In classical mechanics physical quantities Q of a particle other than energy also can be regarded as a function of coordinates $\boldsymbol{r}$ and momentum $\boldsymbol{p}$ of the particle so that Q is expressed as $Q(\boldsymbol{r}, \boldsymbol{p})$. In going to quantum mechanics, both $\boldsymbol{r}$ and $\boldsymbol{p}$ are replaced with the corresponding operators as $\boldsymbol{r} \to \hat{\boldsymbol{r}} = \boldsymbol{r}\times$, i.e., simple multiplication and $\boldsymbol{p} \to \hat{\boldsymbol{p}} = (\hbar/\mathrm{i})(\partial/\partial \boldsymbol{r})$ (cf. (1.17)). Then, Q becomes a function of the operators and is denoted by $\hat{Q}$. An eigenvalue equation for this operator $\hat{Q}$ is written as in (2.7) where f and q are the eigenfunction and eigenvalue, respectively.

$$\hat{Q}f = qf \tag{2.7}$$

Any function f satisfying the boundary conditions is single-valued, finite, and continuous. Such a function is physically acceptable and represents a state, in which the result of measurement of Q coincides with the eigenvalue q.

2.1.4 Operators of Any Physical Quantity are Linear: Quantum Mechanical States are Superposable

The operator $\hat{Q}$ for any physical quantity is required to be a linear operator. This requirement means that the operator must obey the following relations where f_1 and f_2 are any functions to be operated on by $\hat{Q}$ and c is any arbitrary constant (not necessary real).

$$\hat{Q}(f_1 + f_2) = \hat{Q}f_1 + \hat{Q}f_2 \tag{2.8}$$

$$\hat{Q}(cf) = c\hat{Q}f \tag{2.9}$$

The above requirement turns out to be necessary and sufficient in quantum mechanics to explain experimental results of a system of microscopic particles. A state described by any function g, which is a linear combination of functions f_i as $g = \sum c_i f_i$ is called a superposed state with regard to a physical quantity Q. In general, this function g is *not* an eigenfunction of $\hat{Q}$ contrary to the function f in (2.7) (cf. Sect. 2.1.6). However, an exception is the case when all the functions f_i in the above summation happen to have one and the same eigenvalue (cf. Sect. 2.1.7).

Since the above statement is somewhat abstract and intricate, I will explain for the case of the operator $\hat{H}$ in (1.23) reproduced in Sect. 2.1.1, which should be a familiar operator $\hat{Q}$ by now, compared with the other operators in general. We know that the two states describable by the wave functions $\psi_1(\boldsymbol{r}, t)$ $(= f_1)$ and $\psi_2(\boldsymbol{r}, t)$ $(= f_2)$ satisfy the following equations in accordance with (1.23).

$$\begin{aligned} \mathrm{i}\hbar\frac{\partial \psi_1(\boldsymbol{r}, t)}{\partial t} &= \hat{H}\psi_1(\boldsymbol{r}, t) \\ \mathrm{i}\hbar\frac{\partial \psi_2(\boldsymbol{r}, t)}{\partial t} &= \hat{H}\psi_2(\boldsymbol{r}, t) \end{aligned} \tag{2.10}$$

Now, consider a state described by the superposition of the two wave functions as $\psi(\boldsymbol{r},t) = c_1\psi_1(\boldsymbol{r},t) + c_2\psi_2(\boldsymbol{r},t)$, and let the operator $\hat{H}$ act upon this superimposed function. According to the requirements of (2.8) and (2.9), the following relation must hold:

$$\hat{H}\psi(\boldsymbol{r},t) = c_1\hat{H}\psi_1(\boldsymbol{r},t) + c_2\hat{H}\psi_2(\boldsymbol{r},t) \tag{2.11}$$

Substitution of the equations in (2.10) leads to (2.12) below.

$$\begin{aligned}\hat{H}\psi(\boldsymbol{r},t) &= c_1\mathrm{i}\hbar\frac{\partial\psi_1(\boldsymbol{r},t)}{\partial t} + c_2\mathrm{i}\hbar\frac{\partial\psi_2(\boldsymbol{r},t)}{\partial t}\\ &= \mathrm{i}\hbar\frac{\partial}{\partial t}\left[c_1\psi_1(\boldsymbol{r},t) + c_2\psi_2(\boldsymbol{r},t)\right]\\ &= \mathrm{i}\hbar\frac{\partial}{\partial t}\psi(\boldsymbol{r},t)\end{aligned} \tag{2.12}$$

The above equation demonstrates that the superimposed state $\psi(\boldsymbol{r},t)$ also satisfies the relation, $\hat{H}\psi(\boldsymbol{r},t) = \mathrm{i}\hbar\partial\psi(\boldsymbol{r},t)/\partial t$, just as the original two states $\psi_1(\boldsymbol{r},t)$ and $\psi_2(\boldsymbol{r},t)$ do. That is to say, the superimposed state is also allowed in the mechanical system under consideration. In Sects. 2.2.3 and 6.1.2 I will present a concrete example to show that the square of the absolute value of the superimposed wave function $\psi(\boldsymbol{r},t)$ varies with time.

2.1.5 Eigenfunctions are Normalizable

When a function f satisfies the eigenvalue equation $\hat{Q}f = qf$, the following relationship (2.13) also holds by (2.9), which means that both f and its multiple cf equally represent a state associated with an eigenvalue of q.

$$\hat{Q}(cf) = c\hat{Q}f = cqf = q(cf) \tag{2.13}$$

Accordingly, from an arbitrary solution f' of an eigenvalue equation another solution of $f = cf'$ can be obtained in such a way as to satisfy the following relationship:

$$\int |f|^2\mathrm{d}v = c^2\int |f'|^2\mathrm{d}v = 1, \quad c = \pm\frac{1}{\sqrt{\int |f'|^2\mathrm{d}v}} \tag{2.14}$$

The function f is called the normalized eigenfunction and the multiplying constant c is named the normalization constant. Since the sign of the constant does not have a physical significance in most cases, the plus sign is adopted by custom.

2.1.6 Superposition of Non-Degenerate States does not Lead to an Eigenstate

When the eigenvalue equation $\hat{Q}f = qf$ is satisfied by more than one eigenfunction, e.g., f_1 and f_2, and these are associated with the same eigenvalue of $q_1 = q_2$, it is said that the functions f_1 and f_2 are degenerate. Such a case will be considered in Sect. 2.1.7. Here, the superposition of eigenstates for the case of $q_1 \neq q_2$ will be examined first.

A caution is necessary not to be confused with the argument in Sect. 2.1.4; a superimposed state $f = c_1 f_1 + c_2 f_2$ composed of two eigenfunctions f_1 and f_2 associated with different eigenvalues with respect to a physical quantity of Q does not, in general, represent an *eigenstate* of Q. This is because the operation $\hat{Q}$ on the superimposed state merely yields the result $\hat{Q}f = c_1 q_1 f_1 + c_2 q_2 f_2$, which does not satisfy the correct relation of $\hat{Q}f = \text{const} \times f = \text{const} \times (c_1 f_1 + c_2 f_2)$ required for an eigenvalue equation. As an example, I choose as $\hat{Q}$ the Hamiltonian $\hat{H}$ for systems with a time-independent potential in (2.2) (= (2.6)). Then, a superimposed state $c_1 f_1 + c_2 f_2$ composed of two energy eigenstates satisfying the relations $\hat{H} f_1 = E_1 f_1$ and $\hat{H} f_2 = E_2 f_2$ does not represent a state of a certain energy eigenvalue. There is an exceptional case, however, which will be discussed in Sect. 2.1.7 below.

2.1.7 Superposition of Degenerate States does Lead to an Eigenstate

When several states are degenerate, a superimposed state also has the same eigenvalue as those of the constituent degenerate states. Consider that n eigenfunctions $f_{i1}, f_{i2}, \ldots, f_{in}$, have the same eigenvalue (q_i) for an eigenvalue equation of $\hat{Q}f = qf$. Such a system is said to be in n-fold degeneracy, and the following equations hold simultaneously:

$$\hat{Q} f_{i1} = q_i f_{i1}, \quad \hat{Q} f_{i2} = q_i f_{i2}, \ldots, \hat{Q} f_{in} = q_i f_{in} \tag{2.15}$$

In such a case, operation of $\hat{Q}$ upon a linear combination function of $f_i' = \sum_{j=1}^{n} c_j f_{ij}$ gives the following equation, which indicates that the combined function has the same eigenvalue as that of the composite functions.

$$\begin{aligned} \hat{Q} f_i' &= \hat{Q} \sum_{j=1}^{n} (c_j f_{ij}) = \sum_{j=1}^{n} c_j \hat{Q} f_{ij} \\ &= \sum_{j=1}^{n} c_j q_i f_{ij} = q_i \left(\sum_{j=1}^{n} c_j f_{ij} \right) = q_i f_i' \end{aligned} \tag{2.16}$$

Thus, in the n-fold degeneracy case all the original n eigenstates and any linear combination share the same eigenvalue. A concrete example of this case will be presented in Sect. 2.2.6.

2.1.8 Definition of Hermitian Operators

In order for the energy eigenvalue E in the eigenvalue equation (2.2) (= (2.6)), or more generally, q in (2.7) to have physical significance they have to be real, of course. To guarantee this requirement it is necessary to impose the following restriction on any operator in addition to the requirement discussed in Sect. 2.1.4.

$$\int f^* \hat{Q} f \mathrm{d}v = \int f \hat{Q}^* f^* \mathrm{d}v \tag{2.17}$$

The symbol $\hat{Q}^*$ stands for a complex conjugate of $\hat{Q}$, which is obtained by replacing the imaginary unit i everywhere in $\hat{Q}$ with $-$i. The integration is to be carried out over all space of the variables.

Operators satisfying the above condition are called the Hermitian operators. If the function f is an eigenfunction of $\hat{Q}$ satisfying (2.7), the left-hand side of (2.17) may be regarded as an average of eigenvalue of q with the eigenfunction taken as a kind of averaging weight. Since the function f can be assumed to have been normalized already (cf. Sect. 2.1.5), the left-hand side of (2.17) turns out to be:

$$\int f^* \hat{Q} f \mathrm{d}v = q \int f^* f \mathrm{d}v = q \tag{2.18}$$

As for the right-hand side of (2.17), the relationships of $(f^*)^* = f$ and $(\hat{Q}^*)^* = \hat{Q}$ always hold whether f is an eigenfunction of $\hat{Q}$ or not. Therefore, the right-hand side is just the complex conjugate of the left-hand side and vice versa, and the relationship $q = q^*$ is obtained from (2.17) and (2.18). This relationship guarantees that the eigenvalue q is real.

Incidentally, the hermicity of operators is defined more often than not by the following equation (2.19), rather than by (2.17), where the two functions f_1 and f_2 are supposed to be different.

$$\int f_1^* \hat{Q} f_2 \mathrm{d}v = \int f_2 \hat{Q}^* f_1^* \mathrm{d}v \tag{2.19}$$

This perplexing form is a generalization of the definition shown in (2.17). Equation (2.19) appears to be a more stringent requirement than (2.17), but it can be shown, after D. Bohm, that the two definitions are equivalent:

Let $f_1 + f_2$ be f. Substitution of $f = f_1 + f_2$ into (2.17) and rearrangement of each term leads to (2.20).

$$\begin{aligned} &\int \left(f_1^* \hat{Q} f_1 + f_2^* \hat{Q} f_2\right) \mathrm{d}v + \int \left(f_1^* \hat{Q} f_2 + f_2^* \hat{Q} f_1\right) \mathrm{d}v \\ &= \int \left(f_1 \hat{Q}^* f_1^* + f_2 \hat{Q}^* f_2^*\right) \mathrm{d}v + \int \left(f_1 \hat{Q}^* f_2^* + f_2 \hat{Q}^* f_1^*\right) \mathrm{d}v \end{aligned} \tag{2.20}$$

From (2.17) the first terms of both sides of the above equation cancel each other to leave:

$$\int \left(f_1^* \hat{Q} f_2 - f_2 \hat{Q}^* f_1^*\right) \mathrm{d}v = \int \left(f_1 \hat{Q}^* f_2^* - f_2^* \hat{Q} f_1\right) \mathrm{d}v \tag{2.21}$$

Since f_1 and f_2 are taken to be different in general, the equality should hold for another pair of different functions $e^{ia} f_1$ and $e^{ib} f_2$, where a and b are any arbitrary numerals. Substitution of f_1 and f_2 in (2.21) by the latter pair gives (2.22) below.

$$e^{i(b-a)} \int \left(f_1^* \hat{Q} f_2 - f_2 \hat{Q}^* f_1^* \right) dv = e^{i(a-b)} \int \left(f_1 \hat{Q}^* f_2^* - f_2^* \hat{Q} f_1 \right) dv \quad (2.22)$$

Being complex conjugates the phase factors in (2.22) are not one and the same number. In order for (2.22) to be valid despite this difference in the phase factors the integrations on both sides must be identically 0, from which (2.19) is derived. The usefulness of the definition in the form of (2.19) is to be fully recognized in both wave mechanical and matrix mechanical manipulations whenever two different states are involved in the treatment of interstate quantities such as the transition probability and the effect of perturbation.

In Dirac's bracket representation mentioned in Sect. 2.1.1, the generalized definition in (2.19) can be denoted as $\langle f_1|\hat{Q}|f_2\rangle = \langle f_2|\hat{Q}|f_1\rangle^*$ or more concisely, $\langle 1|\hat{Q}|2\rangle = \langle 2|\hat{Q}|1\rangle^*$, which is further abbreviated as $Q_{12} = Q_{21}^*$.

2.1.9 Eigenfunctions of Hermitian Operators are Orthogonal

Eigenfunctions f_m and f_n of a Hermitian operator $\hat{Q}$ corresponding to different eigenvalues q_m and q_n satisfy the equation below where the integration extends to the full range of the variables. Such eigenfunctions are said to be orthogonal.

$$\int f_m^* f_n dv = 0 \quad (2.23)$$

A complementary explanation on orthogonality will be made in Sect. 2.1.13 also.

Problem

2.1 Prove (2.23).
Hint: Take the complex conjugate of the eigenvalue equation of $\hat{Q} f_m = q_m f_m$. Multiply both sides of the resulting equation from the left-hand side with f_n and integrate. In a similar manner multiply, from the left, both sides of another eigenvalue equation of $\hat{Q} f_n = q_n f_n$ with f_m^* and integrate. Take the difference between the two integrations for each side. Then, refer to the definition of hermicity in (2.19).

2.1.10 Eigenfunctions of Hermitian Operators are Complete

Any function describing a system can be expressed by a linear combination of all the eigenfunctions of a certain physical quantity Q of the system. This property is described by saying that eigenfunctions possess completeness. Alternatively, it is stated that any function of the system can be expanded in a series of all the members of the eigenfunctions of Q. The proviso is that the range of the variables of the function under consideration and the boundary conditions for the function must be the same as those of the eigenfunctions to be used for the expansion.

The completeness of eigenfunctions was advocated by J.B. Fourier early in the nineteenth century and was accepted first as a postulate and later regarded as the expansion theorem*. A familiar example of completeness is seen for the Fourier expansion. It is noted that completeness is not unique in its usage to quantum mechanics but that its power is fully exploited in quantum mechanical analysis also.

The above statement is concisely formulated as follows, where g represents an arbitrary function of the system and c_i is an expansion coefficient.

$$g = \sum_i c_i f_i \tag{2.24}$$

The coefficients are given by (2.25) because of the orthogonality of eigenfunctions f_i.

$$\int f_j^* g \mathrm{d}v = \sum_i c_i \int f_j^* f_i \mathrm{d}v = c_j \tag{2.25}$$

From the above equation the function of interest can be expressed also as:

$$g = \sum_i c_i f_i = \sum_i f_i \left(\int f_i^* g \mathrm{d}v \right) \tag{2.26}$$

2.1.11 The Outcome of Observation: Eigenvalues vs. Expectation Values

When the system is in a state described by a function g which is expanded as in (2.24) by eigenfunctions f_i satisfying the eigenvalue equation $\hat{Q}f = qf$, observation of the physical quantity Q yields each time any one of the eigenvalues of $q_1, q_2, \ldots$, randomly. The randomness means that one cannot predict any particular eigenvalue in advance. All that is known is that repeated observation leads to the following average value $\overline{q}$.

$$\begin{aligned}\overline{q} &= \int g^* \hat{Q} g \mathrm{d}v \Big/ \int g^* g \mathrm{d}v \\ &= \int \left(\sum_i c_i^* f_i^*\right) \hat{Q} \left(\sum_j c_j f_j\right) \mathrm{d}v \Big/ \int \left(\sum_i c_i^* f_i^*\right) \left(\sum_j c_j f_j\right) \mathrm{d}v \\ &= \sum_{i,j} c_i^* c_j q_j \int f_i^* f_j \mathrm{d}v \Big/ \sum_{i,j} c_i^* c_j \int f_i^* f_j \mathrm{d}v = \sum_{i,j} c_i^* c_j q_j \delta_{ij} \Big/ \sum_{i,j} c_i^* c_j \delta_{ij} \\ &= \left(\sum_i |c_i|^2 q_i\right) \Big/ \left(\sum_i |c_i|^2\right) \end{aligned} \tag{2.27}$$

This average value is called the expectation value. Note that the functions f are assumed to have been normalized. Also, the following symbol, called Kronecker's delta, is utilized as follows.

$$\delta_{ij} = 1 \ (i = j), \quad \delta_{ij} = 0 \ (i \neq j) \tag{2.28}$$

It is seen from (2.27) that the expectation value is similar to the average of eigenvalues q_i with a weight of $|c_i|^2$.

2.1.12 The Product of Two Operators: Commutators and Simultaneous Eigenfunctions

The product of two operators $\hat{Q}_1$ and $\hat{Q}_2$ is denoted with $\hat{Q}_1\hat{Q}_2$ and is defined by the following equation.

$$\hat{Q}_1\hat{Q}_2 f = \hat{Q}_1(\hat{Q}_2 f) \tag{2.29}$$

The definition implies that the operation of the product of two operators upon a function f is equivalent to the operation of $\hat{Q}_1$ upon the result of the operation of $\hat{Q}_2$ on f. Importantly, the result of the operation of $\hat{Q}_1\hat{Q}_2$ is not always the same as that of $\hat{Q}_2\hat{Q}_1$, i.e., in general, $\hat{Q}_1\hat{Q}_2 f \neq \hat{Q}_2\hat{Q}_1 f$.

When both sides coincide, $\hat{Q}_1$ and $\hat{Q}_2$ are said to be commutative, and the commutator $[\hat{Q}_1, \hat{Q}_2]$ defined as follows becomes 0.

$$[\hat{Q}_1, \hat{Q}_2] \equiv \hat{Q}_1\hat{Q}_2 - \hat{Q}_2\hat{Q}_1 \tag{2.30}$$

In case where $\hat{Q}_1$ and $\hat{Q}_2$ are commutative, there are eigenfunctions common to both $\hat{Q}_1$ and $\hat{Q}_2$, i.e., there are functions which satisfy the eigenvalue equations of both $\hat{Q}_1$ and $\hat{Q}_2$. Such a function is called a simultaneous eigenfunction. In this case, the eigenvalues of both $\hat{Q}_1$ and $\hat{Q}_2$ are definitely determined. Reciprocally, when $\hat{Q}_1$ and $\hat{Q}_2$ share the same series of eigenfunctions, the commutator becomes 0.

The definition of the commutator by (2.30) is made without any specification of functions. Therefore, if one wants to know whether there is a simultaneous eigenfunction or not, one only needs to see whether the difference $\hat{Q}_1\hat{Q}_2 - \hat{Q}_2\hat{Q}_1$ is 0 or not. There is no need to know the explicit form of the function to be operated upon. The weak point of the definition by (2.30), on the other hand, is that the relation of $[\hat{Q}_1, \hat{Q}_2] = 0$ does not necessarily guarantee that *all* the eigenfunctions $\hat{Q}_1$ of coincide with *all* the eigenfunctions of $\hat{Q}_2$.

Since the above description is somewhat abstract, I will illustrate via a practical example. Let us choose the one-dimensional momentum operator $\hat{p} = (\hbar/\mathrm{i})\mathrm{d}/\mathrm{d}x$ of a free particle as $\hat{Q}_1$ and the kinetic energy operator $\hat{H} = (-\hbar^2/2m)\mathrm{d}^2/\mathrm{d}x^2$ of that particle as $\hat{Q}_2$. Since the identity $\mathrm{d}/\mathrm{d}x \times \mathrm{d}^2/\mathrm{d}x^2 - \mathrm{d}^2/\mathrm{d}x^2 \times \mathrm{d}/\mathrm{d}x = 0$ holds always, irrespective of the functions to be operated upon, $\hat{Q}_1$ and $\hat{Q}_2$ are commutative. From the eigenvalue equation of momentum $\hat{p}f = (\hbar/\mathrm{i})\mathrm{d}f/\mathrm{d}x = qf$ it is obvious that the eigenvalue q is dimensionally equivalent to the dimension of mass times velocity, as it should be. Rewriting $q/\hbar \equiv 2\pi k$, the quantity $k(= q/h)$ can be identified with the absolute value of the wave number vector $\boldsymbol{k} = \boldsymbol{p}/h$ discussed in Sect. 1.4.2. The eigenfunction f for the momentum q is easily obtained through simple integration as $f \propto \exp(\mathrm{i}qx/\hbar) = \exp(\mathrm{i}2\pi kx)$. This function represents a plane progressive wave, the direction of the progression being forward or backward. Corresponding to the two directions, the function may be written as $\exp(+\mathrm{i}2\pi kx)$ and $\exp(-\mathrm{i}2\pi kx)$. Because of the commutativity of $\hat{p}$ and $\hat{H}$ these two functions are also eigenfunctions of the kinetic energy operator $\hat{H}$ as can be verified by substituting these functions into the eigenvalue equation of $\hat{H}f = (-\hbar^2/2m)\mathrm{d}^2f/\mathrm{d}x^2 = Ef$ to obtain the result of $E = h^2k^2/(2m)$. This result indicates that the two functions are energetically degenerate so that the linear combinations, $\exp(+\mathrm{i}2\pi kx) \pm \exp(-\mathrm{i}2\pi kx)$, are also the eigenfunctions of $\hat{H}$. These linear combinations represent standing waves in contrast to the two progressive waves (cf. Euler's relation, $\exp(\mathrm{i}\theta) = \cos\theta + \mathrm{i}\sin\theta$). However, these standing wave functions are *not* eigenfunctions of the momentum operator. The weak point mentioned above has, thus, been shown by this example. This rather lengthy explanation was made because a similar situation will be encountered in Sect. 3.3.3.

Problem

2.2 Verify a standing wave function is not an eigenfunction for the momentum operator.

2.1.13 Perturbation Method: Non-Degenerate Case

There are two major approximation methods used in quantum mechanical problems. They are the perturbation method and the calculus of variation. The reason these approximation methods are necessary is that the time-independent Schrödinger equation in (2.6) can seldom be solved exactly. I will explain in brief the perturbation method in Sects. 2.1.13 and 2.1.14, and the calculus of variation in Sect. 2.1.15.

Let us assume that the Hamiltonian $\hat{H}$ of the system under perturbation can be expressed as $\hat{H} = \hat{H}^0 + \hat{H}'$. The operator $\hat{H}^0$ is that for an eigenvalue equation $\hat{H}^0\phi^0 = E^0\phi^0$ of an unperturbed system with both ϕ^0 and E^0 assumed to be known exactly. The difference between $\hat{H}$ and $\hat{H}^0$ denoted as $\hat{H}'$ represents the effect of perturbation on the system relative to the unperturbed state ϕ^0. When the effect of $\hat{H}'$ is small compared with that of $\hat{H}^0$, and when the explicit form of $\hat{H}'$ is known, the sought eigenfunction ϕ and eigenvalue E satisfying the eigenvalue equation of $\hat{H}\phi = E\phi$ can be approximately expressed in terms of the known eigenfunctions ϕ^0 and eigenvalues E^0. This method of approximation is called the perturbation method, which has long been employed in various fields of natural science, e.g., in the analysis of the motion of celestial bodies. In this section I will give a brief explanation of the case in which there is no degeneracy.

A convenient technique to indicate that the effect of the perturbation is small is to adopt a dimensionless scalar λ and assume that the effect of $\lambda\hat{H}'$ on the system is small in comparison with the effect of $\hat{H}^0$. The scalar λ is an artifice used to judge the magnitude of the effect of operators and the magnitude of the values of functions and energy. As soon as that judgement is made, λ is replaced with unity to disappear from our field of view.

Let us start with the unperturbed system described by the eigenvalue equation below.

$$\hat{H}^0\phi_k^0 = E_k^0\phi_k^0 . \tag{2.31}$$

When the system is perturbed, the k-th eigenfunction and the accompanying eigenvalue are assumed to be slightly changed as in the following equations (2.32) and (2.33):

$$\phi_k = \phi_k^0 + \lambda\phi_k^1 + \mathrm{O}(\lambda^2) \tag{2.32}$$

$$E_k = E_k^0 + \lambda E_k^1 + \mathrm{O}(\lambda^2) \tag{2.33}$$

Here, the unknown function ϕ_k and energy E_k are represented by ϕ_k^0 and E_k^0 plus the correctional terms which are assumed to become vanishingly small with the order of scalar λ. The correcting elements ϕ_k^1 and E_k^1 are yet to be determined. The symbol $\mathrm{O}(\lambda^2)$ represents the correction due to all terms of order λ^2 and higher, inclusively. Although there may be a number of terms of higher order for a thorough approximation, I will ignore these higher order terms because this book is intended to give only a basic outline of the approximation procedure.

Although yet unknown, the eigenfunction ϕ_k and the eigenvalue E_k under the perturbation must obey the eigenvalue equation of $\hat{H}\phi_k = E_k\phi_k$. Thus, (2.35) below is obtained by substitution of (2.32) and (2.33).

$$\hat{H}\phi_k = (\hat{H}^0 + \lambda\hat{H}')\phi_k = E_k\phi_k \tag{2.34}$$

$$\hat{H}^0\phi_k^0 + \lambda(\hat{H}^0\phi_k^1 + \hat{H}'\phi_k^0) + \mathrm{O}(\lambda^2) = E_k^0\phi_k^0 + \lambda(E_k^0\phi_k^1 + E_k^1\phi_k^0) + \mathrm{O}(\lambda^2) \tag{2.35}$$

The first terms of both sides of (2.35) cancel out because of (2.31). By equating the terms bearing scalar λ in both sides, the following equation is obtained.

$$\hat{H}^0\phi_k^1 + \hat{H}'\phi_k^0 = E_k^0\phi_k^1 + E_k^1\phi_k^0 \tag{2.36}$$

This equation involves ϕ_k^1 and E_k^1, which are parts of the sought solutions of ϕ_k and E_k as seen from (2.32) and (2.33). The parts of ϕ_k^1 and $\hat{H}'\phi_k^0$ in (2.36) are unknown functions contrary to the original ϕ_k^0. However, these unknown functions should be expressible as linear combinations of the original functions satisfying the eigenvalue equation of (2.31), thanks to the completeness of ϕ_k^0. Let these linear combinations be as follows where c_{jk} and d_{jk} are coefficients which will be determined afterwards.

$$\phi_k^1 = \sum_{j=1} \phi_j^0 c_{jk}, \quad \hat{H}'\phi_k^0 = \sum_{j=1} \phi_j^0 d_{jk} \tag{2.37}$$

Substitution of these equations into (2.36), after rearrangement, gives the following equation.

$$\sum_j \phi_j^0 \left\{c_{jk}(E_j^0 - E_k^0) + d_{jk}\right\} - E_k^1\phi_k^0 = 0 \tag{2.38}$$

The eigenfunctions ϕ_j^0 and ϕ_k^0 of the unperturbed system have the property of orthogonality as stated in Sect. 2.1.9. There, the orthogonality was defined in an abstract way by the integration of (2.23). I will digress a bit here and give another look at the orthogonality to mitigate the abstractness inherent in the definition of (2.23). First, consider the Cartesian coordinate defined by three vectors, $\boldsymbol{x}, \boldsymbol{y}, \boldsymbol{z}$ which are orthogonal in the familiar sense. Mathematically, they are said to be linearly independent, meaning that in order for any arbitrary vector $\boldsymbol{v} = a\boldsymbol{x} + b\boldsymbol{y} + c\boldsymbol{z}$ in the coordinate system to be identically 0, it is mandatory that the relation $a = b = c = 0$ holds. To make doubly sure, consider a simple two dimensional case. If a vector $a\boldsymbol{x} + b\boldsymbol{y}$ in the xy-plane is 0, but if both a and b are not equal to 0, then the basis vectors $\boldsymbol{x}$ and $\boldsymbol{y}$ turn out to be merely collinear and they cannot extend the two dimensional plane. For these vectors to be the real basis both a and b have to be identically 0.

Now, look at the left-hand side of (2.38). The eigenfunctions ϕ_j^0 and ϕ_k^0 should be orthogonal because they are the solution of the eigenvalue equation of (2.31). According to the preceding paragraph, however, they may also be said to be linearly independent. Thus, for the left-hand side to be equal to 0, the coefficients of all the eigenfunctions must be 0. This means that the coefficients for the members of $j \neq k$ shown in the curly brackets must be 0, from which c_{jk} is therefore determined to be $c_{jk} = d_{jk}/(E_k^0 - E_j^0)$. The numerator d_{jk} in $c_{jk} = d_{jk}/(E_k^0 - E_j^0)$ for $j \neq k$ is obtained by $d_{jk} = \int \phi_j^{0*}\hat{H}'\phi_k^0 \mathrm{d}v$ from the second equation of (2.37). Therefore, the first order correction to ϕ_k in (2.32), i.e., ϕ_k^1 is given by the first term of (2.37) as $\phi_k^1 = \sum_{j \neq k} \left\{ \int \phi_j^{0*} \hat{H}' \phi_k^0 \mathrm{d}v / (E_k^0 - E_j^0) \right\} \phi_j^0$.

As for the excepted k-th member, it is required by (2.38) to be $d_{kk} - E_k^1 = 0$. Thus, the first order correction to E_k in (2.33), E_k^1, is given by the second equation of (2.37) as $E_k^1 = \int \phi_k^{0*} \hat{H}' \phi_k^0 \mathrm{d}v$.

In conclusion, for the first order approximation including up to the second terms of (2.32) and (2.33), the perturbed system is described by ϕ_k and E_k, as given by the following formulas:

$$\phi_k \cong \phi_k^0 + \sum_{j \neq k} \frac{\int \phi_j^{0*} \hat{H}' \phi_k^0 \mathrm{d}v}{E_k^0 - E_j^0} \phi_j^0 \tag{2.39}$$

$$E_k \cong E_k^0 + \int \phi_k^{0*} \hat{H}' \phi_k^0 \mathrm{d}v \tag{2.40}$$

The above results show that the approximate energy of the k-th state under perturbation is obtained roughly through an addition of the average of the perturbation effects to the initial energy E_k^0.

As for the eigenfunction, the initial function ϕ_k^0 is modified by the contribution from all the functions $\phi_j^0 (j \neq k)$. The coefficient of each contributing function is given by the ratio of the perturbation strength $\int \phi_j^{0*} \hat{H}' \phi_k^0 \mathrm{d}v$ to the energy difference $E_k^0 - E_j^0$. On account of this factor of the energy difference the perturbation effect from the states close to the k-th state is expected to be generally large.

Higher order approximations than this first order example can be formulated in a similar manner, although the procedure is a little more involved. However, when higher order approximations are necessary, the choice of unperturbed states as the reference states is not quite appropriate. In such a case one should attempt methods of approximation other than the perturbational method. Usually, usage of the perturbation method is limited to second order approximations at most.

It may be that even the first order perturbation approximation is not easy to grasp clearly. Therefore, I will give a concrete example of the application of the perturbation method in Sect. 2.2.7.

2.1.14 Perturbation Method: Degenerate Case

The perturbation method in the presence of degeneracy (cf. Sect. 2.1.6) is a little more intricate than in the absence of degeneracy as discussed in Sect. 2.1.13. In molecular problems there are several cases involving degeneracy. For example, exchange degeneracy is shown to be a key notion for the understanding of chemical bonding in the H_2^+ ion in Sect. 6.1 and in the H_2 molecule in Sect. 7.2.1. Another example of the problem of degeneracy in connection with chemical bonding is encountered in the coordinate bond of transition metals located in a highly symmetric ligand field. The origin of degeneracy is the presence of some geometrical symmetry of the system under consideration. When degeneracy is present, one is confronted with indeterminate states, such as the indeterminacy of the orientation of the nodes shown in Fig. 2.3 in Sect. 2.2.6.

In this section the effect of perturbation applied to the simplest two-fold degenerate case will be explained, continuing to the point of (2.54). Since the results of this section are utilized in Chaps. 6 and 7, the reader is asked to remain patient up to the end of this section.

When the k-th and l-th eigenstates are energetically degenerate, the second term of the right-hand side of (2.39) becomes infinity for the case of $j = l$. Thus, the treatment described in the preceding section breaks down. To circumvent this pitfall it is necessary to reconsider the effect of perturbation from the beginning. For simplicity, it is assumed that the first two members of the unperturbed eigenfunctions ϕ_1^0 and ϕ_2^0 are degenerate, with all the other members being non-degenerate.

As stated in Sect. 2.1.7, in case ϕ_1^0 and ϕ_2^0 are degenerate, any linear combination $c_1\phi_1^0 + c_2\phi_2^0$ also is associated with the eigenvalue of $E_1^0 = E_2^0$. However, there are only two linearly independent combinations among the infinite number of combinations (cf. the explanation in connection with (2.38)).

Since ϕ_1^0 and ϕ_2^0 are normalizable at any time as stated in Sect. 2.1.5, they are assumed to have been normalized. As for the orthogonality discussed in Sect. 2.1.9, ϕ_1^0 and ϕ_2^0 are not guaranteed to be orthogonal a priori. However, it is possible to construct a new set of orthogonal functions in many ways by taking linear combinations of the original functions. Perhaps the most popular method is the Schmidt orthogonalization*.

In this method, out of the pair of ϕ_1^0 and ϕ_2^0, ϕ_1^0 is chosen as the first member of the new orthogonal set, but ϕ_2^0 is replaced with a linear combination $c_1\phi_1^0 + c_2\phi_2^0$ where the coefficients c_1 and c_2 are determined as $c_1 = -S/(1 - S^2)^{1/2}$ and $c_2 = S/(1 - S^2)^{1/2}$ and S is defined as $S = \int \phi_1^{0*}\phi_2^0 \mathrm{d}v$. It should be easily confirmed that this new linear combination is orthogonal to ϕ_1^0 and is normalized. For a third member ϕ_3^0 and on as well, a new orthogonal combination can be achieved by enforcing the condition that they be orthogonal to the preceding orthogonalized members. This procedure can be carried out at any moment because it involves only a simple integration of the product of two functions. Therefore, one can assume that any degenerate functions are already orthogonalized to each other. This assumption simplifies greatly the derivation of the relationships which follow.

After these preliminaries, one can proceed to the examination of the effect of perturbation on degenerate systems in a manner similar to that for non-degenerate systems. First, for the degenerate ϕ_1^0 and ϕ_2^0 a new set is put forth in (2.41) which corresponds to (2.32) for the non-degenerate case.

$$\phi_k = (p_k\phi_1^0 + q_k\phi_2^0) + \lambda\phi_k^1 + \mathrm{O}(\lambda^2) \quad k = 1, 2 \tag{2.41}$$

This equation shows that the two perturbed linearly independent states ϕ_k with $k = 1, 2$ can be approximated by the two linear combinations $p_k\phi_1^0 + q_k\phi_2^0$ along with the correcting function ϕ_k^1 yet to be determined.

Likewise, the energy E_k under the effect of perturbation is assumed by the following equation which corresponds to (2.33) for the case of non-degenerate systems.

$$E_k = E_k^0 + \lambda E_k^1 + \mathrm{O}(\lambda^2) \quad k = 1, 2 \tag{2.42}$$

In contrast to the non-degenerate case, it is necessary to fix the coefficients p_k and q_k in (2.41), in addition to the yet unknown ϕ_k^1 and E_k^1. As for the third state and on, they are assumed to be non-degenerate as stated above so that for these states ϕ_m with $m \geq 3$ the same equations as (2.32) and (2.33) in Sect. 2.1.13 can be used. They are reproduced below.

$$\phi_m = \phi_m^0 + \lambda\phi_m^1 + \mathrm{O}(\lambda^2) \tag{2.43}$$

$$E_m = E_m^0 + \lambda E_m^1 + \mathrm{O}(\lambda^2) \tag{2.44}$$

Now, (2.41) and (2.42) are substituted into (2.34) and the result is expanded to the first order of λ to obtain (2.45) below, which corresponds to the previous (2.35).

$$\begin{aligned} &\hat{H}^0(p_k\phi_1^0 + q_k\phi_2^0) + \lambda\left[\hat{H}^0\phi_k^1 + \hat{H}'(p_k\phi_1^0 + q_k\phi_2^0)\right] \\ &\quad = E_k^0(p_k\phi_1^0 + q_k\phi_2^0) + \lambda\left[E_k^0\phi_k^1 + E_k^1(p_k\phi_1^0 + q_k\phi_2^0)\right] \end{aligned} \tag{2.45}$$

Obviously, the first terms of both sides compensate each other so that by equating the second terms in the square brackets, we are led to the following equation.

$$\hat{H}^0\phi_k^1 + \hat{H}'(p_k\phi_1^0 + q_k\phi_2^0) = E_k^0\phi_k^1 + E_k^1(p_k\phi_1^0 + q_k\phi_2^0) \quad k = 1, 2 \tag{2.46}$$

The unknowns to be determined in the above equation are ϕ_k^1, $\hat{H}'\phi_1^0$, $\hat{H}'\phi_2^0$, E_k^1 *and* the two coefficients p_k and q_k for the cases of $k = 1$ and 2. The procedure to determine the first four unknowns is to use the expansion theorem (cf. Sect. 2.1.10), which should be familiar by now. The first among the four is expanded as follows in accordance with (2.37) in Sect. 2.1.13.

$$\phi_k^1 = \sum_{j=1} \phi_j^0 c_{jk} \tag{2.47}$$

With this expansion the first term of the left-hand side of (2.46) becomes as follows.

$$\hat{H}^0\phi_k^1 = \sum_{j=1} c_{jk}\hat{H}^0\phi_j^0 = \sum_{j=1} c_{jk}E_j^0\phi_j^0 \tag{2.48}$$

The second and the third unknowns in (2.46), i.e., $\hat{H}'\phi_1^0$ and $\hat{H}'\phi_2^0$ are also to be expanded in terms of the unperturbed functions. According to the general expression of the expansion theorem in (2.26), the above two can be explicitly written as follows.

$$\hat{H}'\phi_1^0 = \phi_1^0\left(\int \phi_1^{0*}\hat{H}'\phi_1^0\mathrm{d}v\right) + \phi_2^0\left(\int \phi_2^{0*}\hat{H}'\phi_1^0\mathrm{d}v\right) + \cdots$$
$$+\phi_j^0\left(\int \phi_j^{0*}\hat{H}'\phi_1^0\mathrm{d}v\right) + \cdots \tag{2.49}$$

$$\hat{H}'\phi_2^0 = \phi_1^0\left(\int \phi_1^{0*}\hat{H}'\phi_2^0\mathrm{d}v\right) + \phi_2^0\left(\int \phi_2^{0*}\hat{H}'\phi_2^0\mathrm{d}v\right) + \cdots$$
$$+\phi_j^0\left(\int \phi_j^{0*}\hat{H}'\phi_2^0\mathrm{d}v\right) + \cdots \tag{2.50}$$

Now, the four expansions in (2.47) through (2.50) are substituted at once into (2.46). After a little rearrangement the following equation expressing the linear independence of ϕ_j^0 with $j = 1, 2, 3, \ldots$ is obtained in common to both cases of $k = 1, 2$.

$$\left[p_k\left(\int \phi_1^{0*}\hat{H}'\phi_1^0\mathrm{d}v\right) + q_k\left(\int \phi_1^{0*}\hat{H}'\phi_2^0\mathrm{d}v\right) - p_kE_k^1\right]\phi_1^0$$
$$+\left[p_k\left(\int \phi_2^{0*}\hat{H}'\phi_1^0\mathrm{d}v\right) + q_k\left(\int \phi_2^{0*}\hat{H}'\phi_2^0\mathrm{d}v\right) - q_kE_k^1\right]\phi_2^0$$
$$+\sum_{j=3,4,\ldots}\left[c_j(E_j^0 - E_k^0) + p_k\left(\int \phi_j^{0*}\hat{H}'\phi_1^0\mathrm{d}v\right) + q_k\left(\int \phi_j^{0*}\hat{H}'\phi_2^0\mathrm{d}v\right)\right]\phi_j^0 = 0$$
$$\tag{2.51}$$

The linear dependence of $\phi_{k=1,2}^0$ and $\phi_{j=3,4,\ldots}^0$ requires that all the coefficients in the square brackets in (2.51) must be 0, as already explained in connection with (2.38). From this requirement, the following two 'simultaneous equations' are obtained. The quotation mark is put here because I should like to draw attention to the fact that the two equations involve three rather than two unknowns, i.e., in the present case not only the coefficients p_k and q_k, but also E_k^1 is yet to be determined.

$$p_k\left(\int \phi_1^{0*}\hat{H}'\phi_1^0\mathrm{d}v - E_k^1\right) + q_k\left(\int \phi_1^{0*}\hat{H}'\phi_2^0\mathrm{d}v\right) = 0 \tag{2.52}$$

$$p_k\left(\int \phi_2^{0*}\hat{H}'\phi_1^0\mathrm{d}v\right) + q_k\left(\int \phi_2^{0*}\hat{H}'\phi_2^0\mathrm{d}v - E_k^1\right) = 0 \tag{2.53}$$

Now, for the coefficients p_k and q_k in the above equations to have significant solutions it is necessary and sufficient that the following secular equation holds.

$$\begin{vmatrix} \int \phi_1^{0*}\hat{H}'\phi_1^0\mathrm{d}v - E_k^1 & \int \phi_1^{0*}\hat{H}'\phi_2^0\mathrm{d}v \\ \int \phi_2^{0*}\hat{H}'\phi_1^0\mathrm{d}v & \int \phi_2^{0*}\hat{H}'\phi_2^0\mathrm{d}v - E_k^1 \end{vmatrix} = 0 \tag{2.54}$$

The determinant above gives a quadratic polynomial with regard to E_k^1. Thus, (2.54) yields two solutions. Denoting these solutions by E_1^1 and E_2^1 and substituting them, in turn, into (2.52) and (2.53) the two equations become the simultaneous equations for the two pairs of p_k and q_k. Remember that the initial degenerate pair of ϕ_1^0 and ϕ_2^0 are assumed to be orthonormal. Likewise, the two linearly independent functions $p_1\phi_1^0+q_1\phi_2^0$ and $p_2\phi_1^0+q_2\phi_2^0$ can also be assumed to be normalized (cf. Sect. 2.1.5). Therefore, the restrictive condition $p_k^2+q_k^2=1$ is usable for both cases of $k=1$ and 2, which allows to reach the final concrete result of (p_1, q_1) and (p_2, q_2).

In conclusion, as shown in Fig. 2.1, the first doubly degenerate state is split into two states by the effect of perturbation and the energy values are given, by the first order approximation, as $E_1 = E_1^0 + E_1^1$ and $E_2 = E_2^0 + E_2^1$.

The degree of the approximation of the eigenfunction, on the other hand, remains the same as the degree of the initial pair of degenerate functions. The perturbed functions are given by $\phi_1 = p_1\phi_1^0 + q_1\phi_2^0$ and $\phi_2 = p_2\phi_1^0 + q_2\phi_2^0$.

The effect of the first two degenerate functions upon each of the m-th functions ($m \geq 3$) can be determined by setting the square bracketed coefficients for $\phi_{j=3,4,\ldots}^0$ in (2.51) to be 0. Now that the initially degenerate first two functions are made non-degenerate, the perturbation effect on all the functions can be expressed in accordance with (2.39) and (2.40) in Sect. 2.1.13.

As I mentioned at the beginning of this section, the result of this section will be used in Sect. 6.1 and elsewhere.

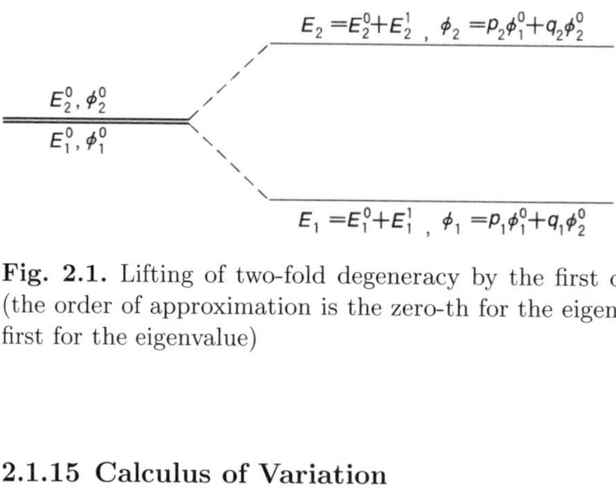

Fig. 2.1. Lifting of two-fold degeneracy by the first order effect of perturbation (the order of approximation is the zero-th for the eigenfunction of energy, and the first for the eigenvalue)

2.1.15 Calculus of Variation

In this section I will give a brief explanation of another approximation method called the calculus of variation. The method is used frequently along with the perturbation method explained in Sects. 2.1.13 and 2.1.14.

Let us consider the eigenvalue equation for energy derived in Sect. 2.1.2 and reproduced below.

$$\hat{H}\phi(\boldsymbol{r}) = E\phi(\boldsymbol{r}) \tag{2.2}$$

Suppose that we are examining various trial functions $\varphi(\boldsymbol{r})$ in search of the exact solution of eigenfunctions $\phi_i(\boldsymbol{r})$ and eigenvalues E_i for the i-th state. When the exact form of the Hamiltonian is known, the expectation value for energy ε_i with respect to the trial function $\varphi(\boldsymbol{r})$ is, in general, different from E_i unless the trial function coincides accidentally with the exact function $\phi(\boldsymbol{r})$. The expectation value ε_i is obtained by the following equation if the trial function is not normalized (cf. Sect. 2.1.11).

$$\varepsilon_i = \int \varphi_i^*(\boldsymbol{r})\hat{H}\varphi_i(\boldsymbol{r})\mathrm{d}v \Big/ \int \varphi_i^*(\boldsymbol{r})\varphi_i(\boldsymbol{r})\mathrm{d}v \tag{2.55}$$

According to the variational principle*, the inequality $\varepsilon_i \geq E_i$ holds for any trial function where the subscript $i = 1, 2, 3, \ldots$ represents the order of increasing values of energy. The inequality relation can be verified straightforwardly for the lowest energy state ($i = 1$) as follows.

Any trial function $\varphi(\boldsymbol{r})$ should be expanded as a linear combination of the exact solution $\phi(\boldsymbol{r})$ by virtue of the expansion theorem (cf. the completeness of eigenfunctions in Sect. 2.1.10). When the trial function is written as $\varphi(\boldsymbol{r}) = \sum_l c_l\phi_l(\boldsymbol{r})$, the problem to solve is the search for the best set of coefficients c_l. Substituting the above linear combination into (2.55) with reference to the orthogonality of $\phi_j(\boldsymbol{r})$ (cf. Sect. 2.1.9) the following equation is obtained.

$$\begin{aligned}\varepsilon_i &= \sum_k\sum_l c_{ki}^* c_{li} \int \phi_k^*\hat{H}\phi_l\mathrm{d}v \Big/ \sum_k\sum_l c_{ki}^* c_{li}\int \phi_k^*\phi_l\mathrm{d}v \\ &= \sum_l |c_{li}|^2 E_l \Big/ \sum_l |c_{li}|^2 \end{aligned}\tag{2.56}$$

Since the minimum among the various energies E_l is set equal to E_1 by assumption, the minimum for the trial function, ε_1, must satisfy the inequality of $\varepsilon_1 \geq E_1$. The proof of the inequality $\varepsilon_i \geq E_i$ for $i \geq 2$ is omitted because it is beyond the level of this book.

So far, the argument is based upon the hypothesis that the trial function $\varphi(\boldsymbol{r})$ could be expanded in terms of yet unknown exact solutions $\phi(\boldsymbol{r})$. Therefore, one cannot proceed any further in finding a concrete form for the trial function. In practice, the trial function is approximated as a finite series of certain known functions $\chi_l(\boldsymbol{r})$ as $\varphi(\boldsymbol{r}) = \sum_l^n c_l\chi_l(\boldsymbol{r})$.

Thus, the problem is how to determine the best coefficients for the minimum energy within the framework of the simple linear combination approximation. This approach is called Ritz's calculus of variation after W. Ritz (1909). Since there is no a priori information on how well the chosen basis functions $\chi_l(\boldsymbol{r})$ will be able to approximate the exact solution nor on how many basis functions will be sufficient for an adequate approximation, one cannot know at the onset the accuracy of the trial functions. With this reservation, however, the best approximate trial functions can be determined by the following procedure.

Substitution of the relation $\phi_l(\boldsymbol{r}) = \sum_l^n c_{li}\chi_l(\boldsymbol{r})$ into (2.55) gives the following expectation value for the i-th state.

$$\varepsilon_i = \sum_k^n \sum_l^n c_{ki}^* c_{li} \int \chi_k^*(\boldsymbol{r})\hat{H}\chi_l(\boldsymbol{r})\mathrm{d}v \Big/ \sum_k^n \sum_l^n c_{ki}^* c_{li} \int \chi_k^*(\boldsymbol{r})\chi_l(\boldsymbol{r})\mathrm{d}v \tag{2.57}$$

For legibility, the integrals $\int \chi_k^*(\boldsymbol{r})\hat{H}\chi_l(\boldsymbol{r})\mathrm{d}v$ and $\int \chi_k^*(\boldsymbol{r})\chi_l(\boldsymbol{r})\mathrm{d}v$ are designated as H_{kl} and S_{kl}, respectively to obtain the expression $\varepsilon_i = \sum_k^n \sum_l^n c_{ki}^* c_{li} H_{kl} / \sum_k^n \sum_l^n c_{ki}^* c_{li} S_{kl}$ and the following equation.

$$\sum_k^n \sum_l^n c_{ki}^* c_{li} H_{kl} - \varepsilon_i \sum_k^n \sum_l^n c_{ki}^* c_{li} S_{kl} = 0 \tag{2.58}$$

Since H_{kl} and S_{kl} are definite integrals in terms of the known functions $\chi_l(\boldsymbol{r})$, they are simply constants. What are sought are the coefficients c_{ki}^* and c_{li}, which give the minimum value for ε_i. Since this value can be viewed as a function of the c_{ki}^* and c_{li}, the first order partial derivatives $\partial\varepsilon_i/\partial c_{ki}^*$ and $\partial\varepsilon_i/\partial c_{li}$ must be 0 for all the coefficients by the necessary condition for extremal values.

Partial differentiation of the left-hand side of (2.58) with regard to each of the coefficients, e.g., c_{ki}^*, yields the following equation,

$$\sum_l^n c_{li} H_{kl} - \varepsilon_i \sum_l^n c_{li} S_{kl} - (\partial\varepsilon_i/\partial c_{ki}^*) \sum_k^n \sum_l^n c_{ki}^* c_{li} S_{kl} = 0\,.$$

From the required conditions, the partial derivative in this equation must be 0. Thus, the first order equation with regard to the c_{li} is obtained as follows,

$$\sum_l^n (H_{kl} - \varepsilon_i S_{kl}) c_{li} = 0\,. \tag{2.59}$$

In the same way, from the differentiation of (2.58) with regard to the coefficient c_{li}, the following equation is obtained,

$$\sum_l^n (H_{lk} - \varepsilon_i S_{lk}) c_{li}^* = 0\,. \tag{2.60}$$

Equations (2.59) and (2.60) appear to be simultaneous equations for the c_{li} and c_{li}^* where the index l runs as $l = 1, 2, \ldots, n$. However, since there is one more unknown, i.e., ε_i in both equations, the equations have $n+1$ unknowns. Therefore, to be more precise, they are 'simultaneous equations' in the same sense as for (2.52) and (2.53). Note that the symbols for the running numbers in (2.60) are changed such that k and l in (2.58) are exchanged for l and k.

In the complex conjugate of (2.60), $\sum_l^n (H_{lk}^* - \varepsilon_i^* S_{lk}^*) c_{li} = 0$, it holds that $H_{kl} = H_{lk}^*$ and $S_{kl} = S_{lk}^*$, because of the hermicity of the Hamiltonian operator and the overlap integrals (cf. Sect. 2.1.8). Therefore, (2.59) and (2.60) are essentially the same so that hereafter only (2.59) will be considered. In order for the set of (2.59) over $k = 1, 2, \ldots, n$ to have significant solutions for the coefficients, it is necessary and sufficient that the following equation be satisfied.

$$\begin{vmatrix} H_{11} - \varepsilon S_{11} & \ldots & H_{1n} - \varepsilon S_{1n} \\ \ldots & \ldots & \ldots \\ H_{n1} - \varepsilon S_{n1} & \ldots & H_{nn} - \varepsilon S_{nn} \end{vmatrix} = 0 \tag{2.61}$$

This is a secular equation similar to the one appearing in Sect. 2.1.14, and is equal to an n-th order equation with respect to ε. There are n real roots which are ordered as $\varepsilon_1 \leq \varepsilon_2 \leq \cdots \leq \varepsilon_n$. When the minimum ε_1 is substituted in (2.59), the ratio of coefficients is obtained as $c_{11} : c_{21} : c_{31} : \cdots : c_{n1}$.

To obtain the absolute values of these coefficients, the additional relationship $\sum_k^n \sum_l^n c_{k1}^* S_{kl} c_{l1} = 1$ is utilized, which is derived from the fact that the trial function, $\varphi_1(\boldsymbol{r}) = \sum_l^n \chi_l(\boldsymbol{r}) c_{l1}$ can be regarded as having been normalized (cf. Sect. 2.1.5). In this way, the best approximate trial function $\varphi_1(\boldsymbol{r})$ and the associated energy ε_1 are fixed.

As in the perturbation method, it is desirable to demonstrate the calculus of variation by also referring to a concrete problem. Such a demonstration will be made in Sect. 2.2.8. This section provides a basis for the arguments to follow in Sects. 6.3.2 and 8.1.

2.2 Application of the Concepts Presented in Sect. 2.1 to the Problem of the Square Well Potential

In this section the fundamentals introduced in Sect. 2.1 will be utilized in solving for the the square well model potential introduced in Chap. 1.

2.2.1 Normalization of Eigenfunctions

Following the procedure in Sect. 2.1.5 the eigenfunctions for the square well potential will be normalized. The function before the normalization is given in (1.4). This function corresponds to f' in (2.14) which represents unnormalized functions in general. The normalization constant c must satisfy the relation, $c^2 \int_{-d}^{d} \sin^2\{n\pi(x + d)/(2d)\}\mathrm{d}x = 1$ by (2.14), from which c is immediately determined to be $d^{-1/2}$ (cf. the identity of $\sin^2\theta = (\cos 2\theta - 1)/2)$). Thus, the normalized function is written as $\phi_n(x) = d^{-1/2}\sin\{n\pi(x + d)/(2d)\}$. In Fig. 1.2 the first three normalized functions are illustrated.

2.2.2 Orthogonality of Eigenfunctions

The general functions f_m and f_n in Sect. 2.1.9 are replaced with the functions, $\phi_m(x) = d^{-1/2}\sin\{m\pi(x + d)/(2d)\}$ and $\phi_n(x) = d^{-1/2}\sin\{n\pi(x + d)/(2d)\}$

with $m \neq n$. The finite integral in (2.23) in the range of $|x| < d$ becomes 0, in reference to the identity $\sin\Theta \sin\Phi = \{\cos(\Theta - \Phi) - \cos(\Theta + \Phi)\}/2$.

2.2.3 Superposition of State: Example 1

As shown in Sect. 2.1.4, superimposed states are also allowed for linear operators such as the time-dependent Hamiltonian. An example of superposition is demonstrated for the first two time-dependent states of the square well potential. Let us assume, for simplicity, that the ground state wave function $\psi_1(\boldsymbol{r}, t)$ and the first excited state wave function $\psi_2(\boldsymbol{r}, t)$ are superimposed at a ratio of 1:1.

As derived in Sect. 2.1.2, the wave function for a system with a time independent potential such as the square well potential can be factored in the form of $\psi(\boldsymbol{r}, t) = \exp(-\mathrm{i}Et/\hbar)\phi(\boldsymbol{r})$. Therefore, for the two states above, the explicit wave functions are written as below,

$$\psi_1(\boldsymbol{r}, t) = \exp(-\mathrm{i}E_1 t/\hbar)\phi_1(\boldsymbol{r}) = d^{-1/2}\exp(-\mathrm{i}E_1 t/\hbar)\sin\{1\pi(x+d)/(2d)\}$$

and

$$\psi_2(\boldsymbol{r}, t) = \exp(-\mathrm{i}E_2 t/\hbar)\phi_2(\boldsymbol{r}) = d^{-1/2}\exp(-\mathrm{i}E_2 t/\hbar)\sin\{2\pi(x+d)/(2d)\}\,.$$

The normalized superposed states $\psi_\pm(\boldsymbol{r}, t)$ with an equal weight of 1:1 are written as $\psi_\pm(\boldsymbol{r}, t) = 2^{-1/2}\{\psi_1(\boldsymbol{r}, t) \pm \psi_2(\boldsymbol{r}, t)\}$, which become at $t = 0$ as follows, $\psi_\pm(\boldsymbol{r}, 0) = 2^{-1/2}\{\psi_1(\boldsymbol{r}, 0) \pm \psi_2(\boldsymbol{r}, 0)\} = (2d)^{-1/2}\sin\{1\pi(x + d)/(2d)\} \pm (2d)^{-1/2}\sin\{2\pi(x + d)/(2d)\}$.

The two functions $\psi_+(\boldsymbol{r}, 0)$ and $\psi_-(\boldsymbol{r}, 0)$ are depicted in solid curves in Figs. 2.2(a) and (b), respectively, along with the constituent functions $\psi_1(\boldsymbol{r}, 0) \equiv \phi_1(\boldsymbol{r})$ and $\psi_2(\boldsymbol{r}, 0) \equiv \phi_2(\boldsymbol{r})$ shown in broken curves.

As to the probability for the superposed wave functions at an arbitrary time t, it is given by $|\psi_\pm(\boldsymbol{r}, t)|^2$ according to Sect. 2.1.1, which is explicitly written with the use of Euler's relation, $\cos\theta = \{\exp(+\mathrm{i}\theta) + \exp(-\mathrm{i}\theta)\}/2$, as below;

$$\begin{aligned}|\psi_\pm(\boldsymbol{r}, t)|^2 &= \psi_\pm^*(\boldsymbol{r}, t)\psi_\pm(\boldsymbol{r}, t)\\ &= \{\exp(+\mathrm{i}E_1 t/\hbar)\phi_1^*(\boldsymbol{r}) \pm \exp(+\mathrm{i}E_2 t/\hbar)\phi_2^*(\boldsymbol{r})\}\\ &\quad \times\{\exp(-\mathrm{i}E_1 t/\hbar)\phi_1(\boldsymbol{r}) \pm \exp(-\mathrm{i}E_2 t/\hbar)\phi_2(\boldsymbol{r})\}/2\\ &= [\phi_1^2(\boldsymbol{r}) + \phi_2^2(\boldsymbol{r}) \pm 2\phi_1(\boldsymbol{r})\phi_2(\boldsymbol{r})\cos\{(E_2 - E_1)t/\hbar\}]/2\end{aligned}$$

Since the cosine term in the above equation varies between +1 and −1, the value of $|\psi_\pm(\boldsymbol{r}, t)|^2$ varies between the two limits, $\{\phi_1(\boldsymbol{r}) + \phi_2(\boldsymbol{r})\}^2/2$ and $\{\phi_1(\boldsymbol{r}) - \phi_2(\boldsymbol{r})\}^2/2$, periodically. From (1.5) in Sect. 1.2 the energy difference in the argument of the cosine in the above equation is calculated as $E_2 - E_1 = (2^2 - 1^2)h^2/(32m_\mathrm{e}d^2) = 3h^2/(32m_\mathrm{e}d^2)$, therefore, as the argument $(E_2 - E_1)t/\hbar = 3ht/(32m_\mathrm{e}d^2)$, becomes equal to $0, 2\pi, 4\pi, \ldots,$ the probability $|\psi_+(\boldsymbol{r}, t)|^2$ becomes $\{\phi_1(x) + \phi_2(x)\}^2/2$. Likewise, at the arguments $1\pi, 3\pi, 5\pi, \ldots,$ the probability is given by $\{\phi_1(x) - \phi_2(x)\}^2/2$.

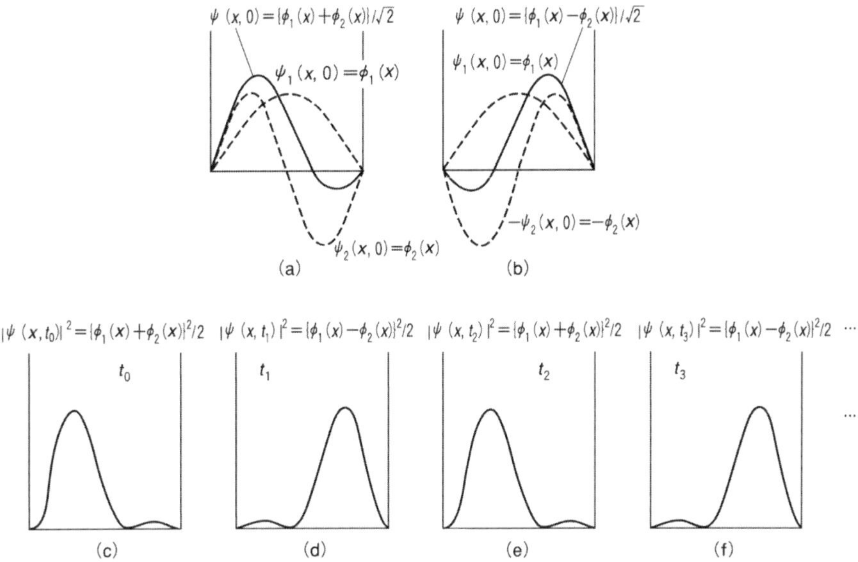

Fig. 2.2. (**a**)–(**b**) Superimposed wavefunctions at $t = 0$ consisting of the first and second states for an electron in the square well potential. (**c**)–(**f**) The time dependence of the probability for the superposed wavefunction $\psi_+(\boldsymbol{r}, t)$. See text for (**a**) through (**f**)

Figures 2.2(c) and (e) correspond to the former case while Figs. 2.2(d) and (f) correspond to the latter case. Thus, in the positively superimposed state $\psi_+(\boldsymbol{r}, t)$, the probability oscillates between the two walls of the potential, which is regarded as a motion of a wave packet. Of course, in the negatively superposed state $\psi_-(\boldsymbol{r}, t)$, the behavior of the probability is similar except for the difference in the phase factor. Conversely, the initial two stationary states can be reproduced by superimposing the two travelling states from the reciprocal relation $\psi_{1,2}(\boldsymbol{r}, t) = 2^{-1/2}\{\psi_+(\boldsymbol{r}, t) \pm \psi_-(\boldsymbol{r}, t)\}$.

The wave packet above represents a sort of interference originating from the superposition of the two states. This wave packet concept will be used in the explanation of the chemical bonding in the H_2^+ ion in Chap. 6. More generally, the wave packet concept is useful to analyze quantum mechanical phenomena dynamically.

2.2.4 Superposition of States: Example 2

In Sect. 2.1.6 it was shown that the linear combination of two eigenfunctions f_1 and f_2 with different eigenvalues does not become a function describing an eigenstate. To confirm this statement for the square well potential let us take $f_1 = d^{-1/2} \sin\{1\pi(x + d)/(2d)\}$ and $f_2 = d^{-1/2} \sin\{2\pi(x + d)/(2d)\}$ and examine, e.g., a 1:1 superposition.

In the case of the square well potential the operation of the Hamiltonian upon f_1+f_2 is explicitly written as $\hat{H}(f_1+f_2)=\{-\hbar^2/(2m_\mathrm{e})\}(\mathrm{d}^2/\mathrm{d}x^2)(f_1+f_2)$, from which the following result is immediately obtained as follows: $\{h^2/(32m_\mathrm{e}d^2)\}(1^2d^{-1/2})\sin\{1\pi(x+d)/(2d)\}+\{h^2/(32m_\mathrm{e}d^2)\}(2^2d^{-1/2})\sin\{2\pi(x+d)/(2d)\}$.

This result demonstrates that the initial ratio of 1:1 for the weights of f_1 and f_2 is changed to a different ratio of 1:4 after the operation. Thus, the unnormalized 1:1 superimposed state cannot be an eigenstate.

2.2.5 Expectation Value of the Energy

As shown above the 1:1 superimposed state does not give a correct eigenvalue. Instead, however, an expectation value can be obtained as stated in Sect. 2.1.11. The normalized 1:1 state is described by $2^{-1/2}(f_1+f_2)$, which corresponds to the general function g in Sect. 2.1.11. Regarding the general operator $\hat{Q}$ in Sect. 2.1.11 as corresponding to the Hamiltonian $\hat{H}$, one can obtain the expectation value of energy $\overline{E}$ according to (2.27), which turns out to be $\overline{E}=h^2/(32m_\mathrm{e}d^2)\times 5/2$. The result can be interpreted as follows. If one repeats the observation of energy in the 1:1 superimposed state, the result obtained upon each observation is either $E_1=h^2/(32m_\mathrm{e}d^2)\times 1^2$ or $E_2=h^2/(32m_\mathrm{e}d^2)\times 2^2$, and through a large number of observations, the result will converge to the above expectation value, which corresponds to an average.

2.2.6 Superposition of States: Example 3

In the example above, the initial two states f_1 and f_2 were not degenerate. However, as discussed in Sect. 2.1.7, superposition of degenerate states gives the same eigenvalue as the initial degenerate states. As an example, let us consider the case of a two-dimensional square well potential with the two wall-to-wall distances being equal to $2d$. It should be obvious that the eigenfunctions are given as $\phi(x,y)=d^{-1}\sin\{n_x\pi(x+d)/(2d)\}\sin\{n_y\pi(y+d)/(2d)\}$ and the eigenvalues are as $E=(n_x^2+n_y^2)h^2/(32m_\mathrm{e}d^2)$, where n_x and n_y run as $n_x,n_y=1,2,3,\ldots$ and the ground state corresponds to $n_x=n_y=1$ with an energy of $E=2h^2/(32m_\mathrm{e}d^2)$, while the first excited state is doubly degenerate with $n_x=2$, $n_y=1$ and $n_x=1$, $n_y=2$ and with the energy $5h^2/(32m_\mathrm{e}d^2)$.

The two degenerate eigenfunctions denoted as ϕ_{2x} and ϕ_{2y} are already orthogonal because of the relation $n_x\neq n_y$. As stated in Sect. 2.1.7, any linear combination $c_1\phi_{2x}+c_2\phi_{2y}$ is also degenerate to ϕ_{2x} and ϕ_{2y} with the same energy. Therefore, there are infinite eigenfunctions as the ratio of $c_1:c_2$ varies continuously.

However, there are only two linearly independent functions for the case of the dual degeneracy. These two are expressed in general as $\phi'_{2x}=c_{11}\phi_{2x}+$

$c_{21}\phi_{2y}$ and $\phi'_{2y} = c_{12}\phi_{2x} + c_{22}\phi_{2y}$. In the matrix representation they may be written as,

$$(\phi'_{2x}, \phi'_{2y}) = (\phi_{2x}, \phi_{2y}) \begin{pmatrix} c_{11} & c_{12} \\ c_{21} & c_{22} \end{pmatrix} .$$

They may also be compactly rewritten as $\boldsymbol{\phi}' = \boldsymbol{\phi c}$ with the vectors and the matrix being denoted in bold face. The matrix $\boldsymbol{c}$ and its inverse matrix $\boldsymbol{c}^{-1}$ are related to each other by the equation $\boldsymbol{cc}^{-1} = \boldsymbol{c}^{-1}\boldsymbol{c} = \boldsymbol{E}$, where $\boldsymbol{E}$ stands for the unit matrix in which all the diagonal elements are unity and all the off-diagonals are 0. From this equation, one can obtain the relationship $\boldsymbol{\phi} = \boldsymbol{\phi}'\boldsymbol{c}^{-1}$, so that one may choose ϕ'_{2x} and ϕ'_{2y} as a substitute set of the linearly independent functions in place of the original set ϕ_{2x} and ϕ_{2y}. If the matrix $\begin{pmatrix} \cos\theta & \sin\theta \\ -\sin\theta & \cos\theta \end{pmatrix}$ is chosen as $\boldsymbol{c}$, the set ϕ'_{2x} and ϕ'_{2y} is obtained by rotating the set ϕ_{2x} and ϕ_{2y} by an angle of θ around the z-axis as shown in Fig. 2.3.

The two components of this new set, of course, have the orthogonality stated in Sect. 2.1.9. Figure 2.3 demonstrates semiquantitatively the amplitudes of ϕ_{2x}, ϕ_{2y} and ϕ'_{2x} and ϕ'_{2y} as a contour map along the direction

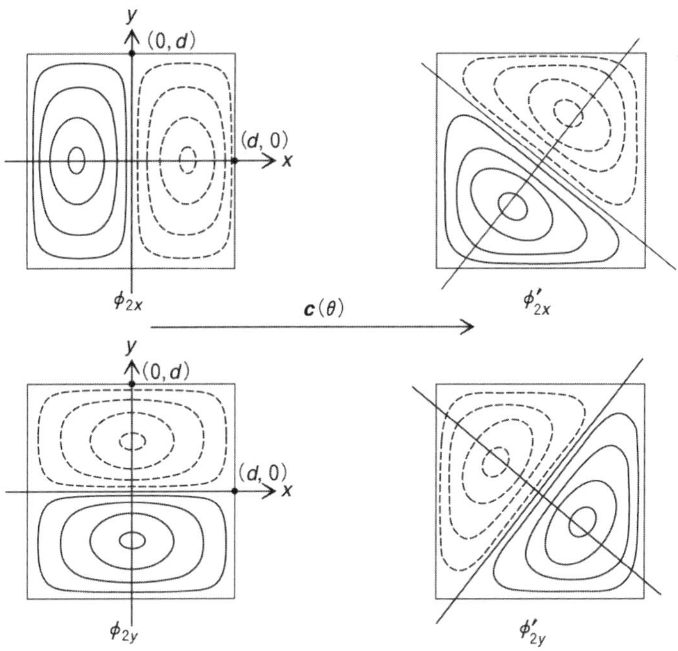

Fig. 2.3. Contour maps to demonstrate that the doubly degenerate set for a two-dimensional square well potential exists infinitely with a continuous change of the rotation angle

perpendicular to the xy-plane. The solid and the dotted lines indicate that the amplitudes are above and below the xy-plane, respectively. Both columns of the figure indicate that the sets ϕ_{2x}, ϕ_{2y} and ϕ'_{2x}, ϕ'_{2y}, are inseparable, and by using these independent components the first excited state is described over the full range of $|x| < d$ and $|y| < d$. It is noted that the dimensions of the amplitudes are equal to the reciprocal of distance in accordance with the statement in Sect. 2.1.1.

Situations similar to that in Fig. 2.3 will be encountered in the discussion of molecular orbitals of ammonia in Sect. 8.2.2. Another popular example is seen for the doubly degenerate electronic states of benzene.

2.2.7 An Example of the Perturbation Method: A Non-Degenerate Case

In Sect. 1.1, I tried to describe in approximate terms an electron in liquid ammonia via a square well potential. If we consider this system a little more realistically, we notice that the electron may be influenced by an alkali metal positive ion which should coexist with the electron. In this section I will examine the effect of the alkali metal ion by the perturbation method in Sect. 2.1.13.

Figure 2.4(a) is supposed to approximate the potential felt by the electron when a Na^+ ion approaches from the left-hand side. The electron will be in an electric field (ε), which is assumed to provide an additional potential $e\varepsilon x$. In this new potential field the electron will be slightly inclined toward the ion. Such a state should be described by a new function rather than the simple trigonometric functions for the unperturbed system. When the effect of the ion is small, one may use the perturbation method in Sect. 2.1.13.

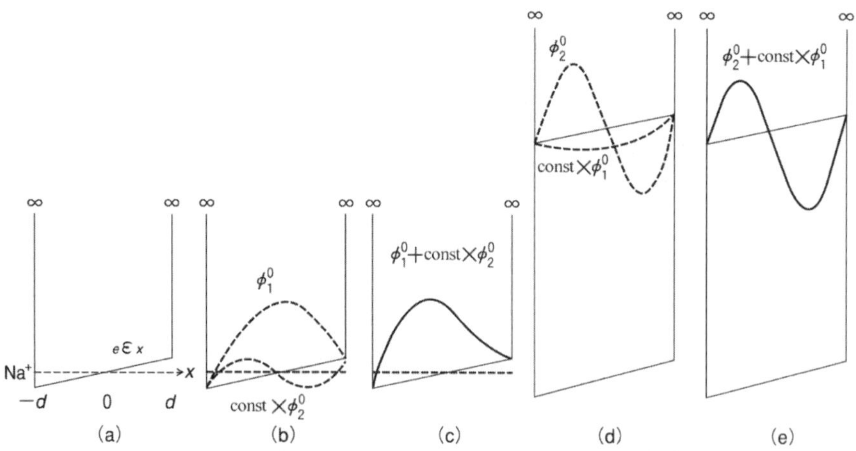

Fig. 2.4. One-dimensional square well potential modified slightly by the effect of a Na^+ ion and the perturbed first (**c**) and second (**e**) electronic states of electron

In the present system, the unperturbed Hamiltonian $\hat{H}^0$ and the perturbing Hamiltonian $\hat{H}'$ are, respectively, given as $\hat{H}^0 = -\hbar^2/(2m_\mathrm{e}) \times \mathrm{d}^2/\mathrm{d}x^2$ and $H' = e\varepsilon x$. The eigenfunctions in the absence of the perturbation, i.e., $\phi_k^0(x) = d^{-1/2}\sin\{k\pi(x+d)/(2d)\}$ and the associated eigenvalues $E_k^0 = h^2k^2/(32m_\mathrm{e}d^2)$ are all known explicitly. Therefore, the integrals in (2.39) and (2.40) are obtained immediately. As a result, the second terms of the right-hand side of (2.39) and (2.40) are given, respectively, as follows,

$$\sum_{j\neq k} d^{-1}\left[\int \sin\{j\pi(x+d)/(2d)\}e\varepsilon x \sin\{k\pi(x+d)/(2d)\}\mathrm{d}x\right]$$
$$\times \sin\{j\pi(x+d)/(2d)\}/(E_k^0 - E_j^0)$$

and

$$d^{-1}\int \sin\{k\pi(x+d)/(2d)\}e\varepsilon x \sin\{k\pi(x+d)/(2d)\}\mathrm{d}x.$$

The second integral above is 0 because the integrand is an odd function. In other words, to the first order approximation, the eigenvalue of energy remains unaffected. If the order of the approximation is elevated, it is ensured from the second order approximation of the perturbation method that the energy of the ground state be slightly stabilized.

As for the ground state function, the first order approximation modifies the original eigenfunction to $\phi_1 \cong \phi_1^0 + \sum_{j\neq 1}\left[\int \phi_j^{0*}\hat{H}'\phi_1^0\mathrm{d}v\right]\phi_j^0/(E_1^0 - E_j^0)$. For simplicity, the contribution from the excited states to the ground state is limited to only the nearest first excited state ($j = 2$), which is reasonable because of the large factor of $1/(E_1^0 - E_{j=2}^0)$. Thus, the ground state function is approximated as,

$$\phi_1 \approx \phi_1^0 + \left[d^{-1}\int \sin\{2\pi(x+d)/(2d)\}e\varepsilon x \sin\{1\pi(x+d)/(2d)\}\mathrm{d}x\right]\phi_2^0/(E_1^0 - E_2^0).$$

The finite integral can be obtained as $-32e\varepsilon d/(9\pi^2) < 0$ through integration by parts (the identity $\cos^3\theta = (3\cos\theta + \cos 3\theta)/4$ may be made use of). Since the denominator $E_1^0 - E_{j=2}^0 = (1^2 - 2^2)h^2/(32m_\mathrm{e}d^2)$ is also negative, ϕ_1 can be written as $\phi_1 \approx \phi_1^0 + \mathrm{const}(>0) \times \phi_2^0$. This shows that the ground state under the perturbation is modified by the additional contribution of ϕ_2^0 and the electron is shifted toward the left as expected. Figure 2.4(c) demonstrates qualitatively such a modification.

Similarly, the perturbed first excited state is approximated as below,

$$\phi_2 \approx \phi_2^0 + \left[d^{-1}\int \sin\{1\pi(x+d)/(2d)\}e\varepsilon x \sin\{2\pi(x+d)/(2d)\}\mathrm{d}x\right]\phi_1^0/(E_2^0 - E_1^0).$$

In this case, while the finite integral is $-32e\varepsilon d/(9\pi^2) < 0$ as before, the denominator is $E_2^0 - E_1^0 > 0$. Thus, ϕ_2 is written as $\phi_2 \approx \phi_2^0 + \mathrm{const}(<0) \times \phi_1^0$. Therefore, the relative signs of ϕ_1^0 and ϕ_2^0 are reversed as shown in Fig. 2.4(d), in contrast to Fig. 2.4(b), and the electron in this excited state tends to move away from the ion as depicted schematically in Fig. 2.4(e).

2.2.8 An Example of the Calculus of Variation: A Non-Degenerate Case

Let us practice the calculation of variation by referring to the above example of liquid ammonia containing an electron and an alkali metal ion. Ritz's method (cf. Sect. 2.1.15) will be used. As a trial function $\varphi(\boldsymbol{r})$, the linear combination $\varphi(\boldsymbol{r}) = c_1\phi_1^0 + c_2\phi_2^0$ is examined. According to the procedure in Sect. 2.1.15, the variationally best energy and function are obtained by starting with the following secular equation:

$$\begin{vmatrix} H_{11} - \varepsilon S_{11} & H_{12} - \varepsilon S_{12} \\ H_{21} - \varepsilon S_{21} & H_{22} - \varepsilon S_{22} \end{vmatrix} = 0$$

Since the unperturbed eigenfunctions ϕ_1^0 and ϕ_2^0 are orthonormalized, the overlap integrals S_{11} and S_{22} are 1 while S_{12} and S_{21} are 0. The remaining integrals in the above determinant are given as follows,

$$\begin{aligned} H_{kk} &= \int \phi_k^{0*}(\hat{H}^0 + \hat{H}')\phi_k^0 \mathrm{d}v = E_k^0 + \int \phi_k^{0*}\hat{H}'\phi_k^0 \mathrm{d}v \\ &= E_k^0 + d^{-1}\int \sin\{k\pi(x+d)/(2d)\}e\varepsilon x \sin\{k\pi(x+d)/(2d)\}\mathrm{d}x \end{aligned}$$

with $k = 1, 2$ and

$$H_{12} = H_{21} = \int \phi_1^{0*}(\hat{H}^0 + \hat{H}')\phi_2^0 \mathrm{d}v = \int \phi_1^{0*}\hat{H}'\phi_2^0 \mathrm{d}v\,.$$

The second term of H_{kk} is 0 because the integrand is an odd function while the last integral $H_{12} = H_{21}$ is $-32e\varepsilon d/(9\pi^2)$ as is shown in Sect. 2.2.7. Inserting $-32e\varepsilon d/(9\pi^2) = -A$, the above secular equation is written as below,

$$\begin{vmatrix} E_1^0 - \varepsilon & -A \\ -A & E_2^0 - \varepsilon \end{vmatrix} = 0\,.$$

This is a quadratic equation with regard to ε, and the two roots, ε_1 and ε_2 are approximated as $\varepsilon_1 \cong E_1^0 - A^2/(E_2^0 - E_1^0)$ and $\varepsilon_2 \cong E_2^0 + A^2/(E_2^0 - E_1^0)$ when $A \ll E_2^0 - E_1^0$.

In order to see the plausibility of the above treatment let us consider a case where the parameters are chosen to mimic the realistic situation. If the perturbation Hamiltonian $\hat{H}' = e\varepsilon x$ is assumed to be 0.1 eV (1 eV = 0.16022×10^{-18} J) at $x = d$ where d is chosen as $d \approx 0.8$ nm in conformity with the examination in Sect. 1.2, the off-diagonal element A in the above determinant becomes $A = 32e\varepsilon d/(9\pi^2) \cong 36$ meV. At $d \approx 0.8$ nm the diagonal elements are calculated as $E_1^0 \cong 1^2h^2/(32m_\mathrm{e}d^2) \cong 146$ meV and $E_2^0 \cong 2^2h^2/(32m_\mathrm{e}d^2) \cong 585$ meV. Meanwhile, the solutions of the above quadratic equations are found to be $\varepsilon \cong 143$ meV and $\varepsilon_2 \cong 588$ meV. Thus, it is known that the perturbation causes a stabilization of about 3 meV for the ground state and a destabilization by about the same amount for the first excited state.

As for the perturbed functions, the ratios of the combination coefficients are calculated by the procedure shown in Sect. 2.1.15 as $c_{11} \approx +20c_{12}$ for the ground state and $c_{21} \approx -c_{22}/20$ for the first excited state. Therefore, by the calculus of variation also, the energy shift and the distortion of the function are predicted in a similar fashion to the results produced by the perturbation method.

Figure 2.5 shows schematically the result of the above calculation. Note that the distortion of the eigenfunctions in Fig. 2.5 is exaggerated.

In this section several fundamental concepts from Sect. 2.1 have been demonstrated by utilizing the fact that the square well model potential allows for the exact solution of the eigenvalue equation of energy of the electron. However, such an exact solution cannot be expected for most problems involving atoms and molecules, except for the one-electron problem of the H atom (Chap. 3) and of the H_2^+ ion (Chap. 6). Since the exceptional case of the one-electron atom where the electron is in a Coulombic potential is rigorously solved, one can make full use of this system as a reference to get approximate solutions for many-electron atoms (Chap. 4) and molecules (Chap. 6 and on).

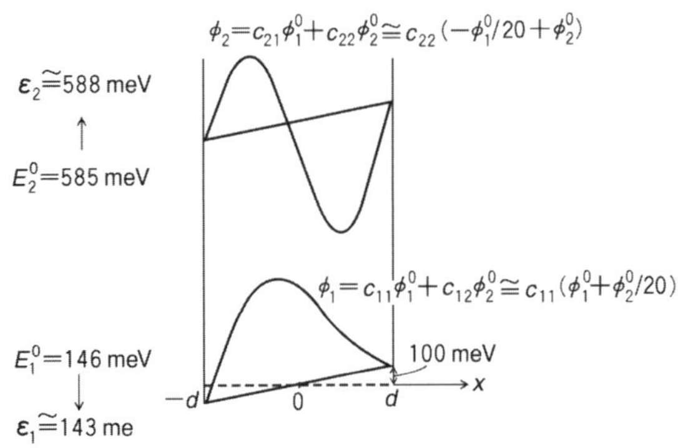

Fig. 2.5. Approximate solutions of the same system as in Fig. 2.4 treated by the calculus of variation

3 One-Electron Atoms: The Fundamental System

As described in Sect. 1.1, the energy state of an electron is determined by the surrounding electrostatic potential. In this chapter, the energy and the angular momentum of a single electron in a Coulombic potential due to the positively charged nucleus, will be studied. This Coulombic potential is not an artificial potential, such as the square well potential introduced in Chap. 1, but is a real potential felt by the electron in a one-electron atom. Therefore, when the eigenvalue equation of energy is solved, the eigenvalues obtained should agree with experimentally observed values within the non-relativistic limit. As for the eigenfunction associated with the exact eigenvalue, the orthogonality and completeness inherent in the function are powerful properties which enable us to get approximate solutions for the complicated systems which will be dealt with in Chap. 4 and afterward. Fortunately, the eigenfunction for systems in the Coulombic potential was thoroughly investigated a couple of centuries ahead of the advent of quantum mechanics, and we can fully avail ourselves of this fruit from the classics.

Today, we understand the periodicity of elements quite thoroughly in terms of the electronic shell structure of atoms. Such insight into the structure of the atom is based entirely on the success in obtaining the exact solution for one-electron atom problems (Chap. 4). As stated in Chap. 1, the H atom has provided the momentum for the great jump from classical to quantum mechanics by way of its line spectrum. In addition, the H atom turns out to be the fundamental basis for quantum mechanical analyses of matter.

In this section we encounter mathematical expressions which may appear complicated. However, it is not necessary to dig into the details of their derivation, rather, what is important is to grasp the physical concepts and insights imparted by these mathematical expressions.

3.1 The Derivation of Energy and Orbital Angular Momentum of the Electron in a Coulombic Potential: An Outline
This section summarizes the rather lengthy derivation of various relations used in the subsequent sections.

3.2 The Derivation of the Eigenfunctions and Eigenvalues of Energy and Orbital Angular Momentum of a Single Electron in a Coulombic Potential

The reader is asked to pay attention to how naturally the eigenfunctions and eigenvalues are derived by imposing the boundary conditions.

3.3 Some Comments on the Solution for One-Electron Atoms
Some comments will be made to assist the reader in understanding the results determined in Sect. 3.2.

3.4 Angular Momentum in One-Electron Atoms
Electron spin is derived by generalizing the definition for the orbital angular momentum.

3.1 The Derivation of Energy and Orbital Angular Momentum of the Electron in a Coulombic Potential: An Outline

Detailed procedures to derive the energy and orbital angular momentum from the Schrödinger equation will be elaborated on in Sect. 3.2. In this section, major conclusions to be reached in the next section will be preliminarily summarized, along with a brief look at the derivation. This redundant presentation is deliberate and intended to prevent the reader from becoming lost while attempting to grasp the logical flow in Sect. 3.2 through frustration with the intervening mathematical formulas.

The one-electron atom in the title includes such systems as the H atom with one proton and one electron, the mono-positive ion of He^{1+} with one bare helium ion He^{2+} and one electron, and the bi-positive ion of Li^{2+} with one bare lithium ion Li^{3+} and one electron. In general, it is a two-particle system consisting of one bare nucleus A^{Z+} and one electron where A is the symbol of atom A and Z stands for the atomic number. Since the mass of the nucleus is more than three orders of magnitude larger than the mass of the electron, the nucleus is assumed to be fixed at a certain point of space. A note addressing this assumption is located at the end of this section.

In the one-electron atom the electron feels a Coulombic potential $V = -Ze^2/(4\pi\varepsilon_0 r)$ where r and ε_0 stand for the distance between the electron and nucleus and the permittivity of a vacuum, respectively. Since this potential is time-independent, the wave function $\psi(\boldsymbol{r}, t)$ can be factored by separation of variables as $\psi(\boldsymbol{r}, t) = \exp(-\mathrm{i}Et/\hbar)\phi(\boldsymbol{r})$, in accordance with the explanation in Sect. 2.1.2. The eigenvalue equation for $\phi(\boldsymbol{r})$ is $\hat{H}\phi(\boldsymbol{r}) = E\phi(\boldsymbol{r})$ with $\hat{H} \equiv \hbar^2\Delta/(2m) - Ze^2/(4\pi\varepsilon_0 r)$ (cf. (3.6)). Note that the symbol of m_e used so far has been replaced with m for the reason stated in the note at the end of this section. The range of the variable $\boldsymbol{r}$ in the above equation is from the origin (the position of the nucleus) to infinity, i.e., the whole of space. The eigenfunction obeying this equation must satisfy the boundary conditions, or equivalently, the function must not diverge for both approaches of $r \to 0$ and $r \to \infty$. Otherwise, a sensible physical meaning cannot be attributed to the function. The procedure to extract only non-divergent solutions from the family of mathematically permissible general solutions is similar to the procedure to select solutions satisfying the boundary conditions considered in the example from Sect. 1.2 for the square well potential.

It will be shown that the search for functions satisfying the above stated boundary conditions yields automatically the eigenvalues of energy for the one-electron atom as $E = -Z^2me^4/\{(4\pi\varepsilon_0\hbar)^2 \times 2n^2\}$. Here, the symbol n assumes the values of $1, 2, 3, \ldots, \infty$. The above result is in perfect agreement with the result in (1.11), which was obtained by Bohr for the hydrogen atom with $Z = 1$ on the basis of the old quantum mechanics.

Besides the energy, the orbital angular momentum of the electron in the one-electron atom is also rigorously determined by the fact that the Coulombic potential is spherically symmetrical. Again, by the boundary conditions proper to the angular motion of the electron, the eigenvalue of orbital angular momentum is restricted to $\sqrt{l(l+1)}\hbar$ where l can assume the values of $0, 1, 2, 3, \ldots, (n-1)$ exclusively. The orbital angular momentum is a vectorial quantity having three independent components parallel to the Cartesian coordinate axes. It will be shown in the next section that not only the orbital angular momentum itself but also one of the three components must be restricted, this time to $m\hbar$ where m runs over $0, \pm1, \pm2, \pm3, \ldots, \pm l$ only. Again, this restriction originates from the boundary conditions.

As for the eigenfunction, it is generally expressed by a product of two factors. One factor depends only on the centerline distance r between the nucleus and the electron. This factor is a polynomial of r. The other factor is purely an angular function of two variables θ and φ. The variables specify the orientation of the electron relative to the nucleus. These variables are comparable with the polar angle from the north pole and the azimuthal angle from a fixed point on the equatorial line. The first radial function is a solution of a confluent hypergeometric differential equation*. The second angular function is from another set of special functions called the spherical surface harmonics. The concrete forms of both functions will be given in Sect. 3.2. Although detailed derivation of these functions is not necessary for the reader, it is hoped that the reader is aware of the general features of the functions and the origin of the quantum numbers which are introduced automatically by the necessity of meaningful physical significance.

Note: The kinetic energy of one-electron atoms is the sum $(1/2)M\dot{\boldsymbol{r}}_{\mathrm{N}}^2 + (1/2)m_{\mathrm{e}}\dot{\boldsymbol{r}}_{\mathrm{e}}^2$ according to the expression from classical mechanics, where $\boldsymbol{r}_{\mathrm{N}}$ and $\boldsymbol{r}_{\mathrm{e}}$ are the coordinates of the nucleus and the electron, respectively, from a common origin. The above sum can be rewritten identically as $(1/2)(M\dot{\boldsymbol{r}}_{\mathrm{N}} + m_{\mathrm{e}}\dot{\boldsymbol{r}}_{\mathrm{e}})^2/(M + m_{\mathrm{e}}) + (1/2)Mm_{\mathrm{e}}(\dot{\boldsymbol{r}}_{\mathrm{N}} - \dot{\boldsymbol{r}}_{\mathrm{e}})^2/(M + m_{\mathrm{e}})$. The sum of $M\dot{\boldsymbol{r}}_{\mathrm{N}} + m_{\mathrm{e}}\dot{\boldsymbol{r}}_{\mathrm{e}}$ in the first term represents the momentum of the atom as a whole, which is written as $\boldsymbol{P}$. Therefore, the first term corresponds to the kinetic energy of the atom as a whole, while the second term represents the internal kinetic energy of the system. Only the latter term is of concern here, which is interpreted as representing the kinetic energy of a virtual particle whose velocity is equal to the relative velocity $\dot{\boldsymbol{r}} \equiv \dot{\boldsymbol{r}}_{\mathrm{e}} - \dot{\boldsymbol{r}}_{\mathrm{N}}$ and whose effective mass is equal to $m \equiv Mm_e/(M + m_e)$. Writing the momentum of this relative motion as $\boldsymbol{p} = m(\dot{\boldsymbol{r}}_e - \dot{\boldsymbol{r}}_N)$, the internal kinetic energy is written as $(1/2)\boldsymbol{p}^2/m$. When $M \gg m_e$, m is approximated as $m \cong m_e(1 - m_e/M) \approx m_e$. Although the effective mass m is slightly reduced in comparison with m_e, it is close enough to the mass of the electron. Therefore, the relative motion of the virtual particle represents the motion of an electron around a sufficiently heavy nucleus.

Problem

3.1 Let the mass of a proton, deuteron, and electron be $m_p = 1.67262 \times 10^{-27}$ kg, $m_d = 3.34250 \times 10^{-27}$ kg and $m_e = 9.10938 \times 10^{-31}$ kg, respectively. Supposing that the hydrogen atom H and deuterium atom D are one-electron atoms consisting of a proton and an electron and of a deuteron and an electron, calculate the effective masses m for these atoms.

3.2 The Derivation of the Eigenfunctions and Eigenvalues of Energy and Orbital Angular Momentum for a Single Electron in a Coulombic Potential

3.2.1 A General Discussion of the Schrödinger Equation

The Schrödinger equation for one-electron atoms is given as follows:

$$\left[-\frac{\hbar^2}{2m}\left(\frac{\partial^2}{\partial x^2} + \frac{\partial^2}{\partial y^2} + \frac{\partial^2}{\partial z^2}\right) + V(r)\right]\phi(\boldsymbol{r}) = E\phi(\boldsymbol{r})$$
$$V(r) = -Ze^2/(4\pi\varepsilon_0 r) \tag{3.1}$$

In order to solve equations such as the above, it is of the utmost importance to choose the most appropriate coordinate system. In the present case, the relative motion of the electron to the nucleus is described by a virtual particle as explained in the note at the end of Sect. 3.1. Since this particle, with the effective mass of m, is in a spherically symmetrical potential $V(r)$, the coordinate system most appropriate for the particle is not the usual Cartesian coordinate system $\boldsymbol{r} = (x, y, z)$, but the polar coordinate system, which is based on the radius $r (\equiv |\boldsymbol{r}_e - \boldsymbol{r}_N|)$, the polar angle θ, and the azimuthal angle φ, as mentioned in the preceding section (cf. Fig. 3.1). In the following, this particle is called an 'electron' in light of the explanation in the preceding note.

The new coordinate system, called the polar coordinate system, is related uniquely to the rectangular coordinate system (the Cartesian system) x, y, z as $x = r \sin\theta \cos\varphi$, $y = r \sin\theta \sin\varphi$, and $z = r \cos\theta$.

On the basis of this correspondence, the Laplacian in the left-hand side of (3.1), introduced previously in (3.21), can be expressed as in (3.2) below after a lengthy and tedious series of manipulations, which are omitted here.

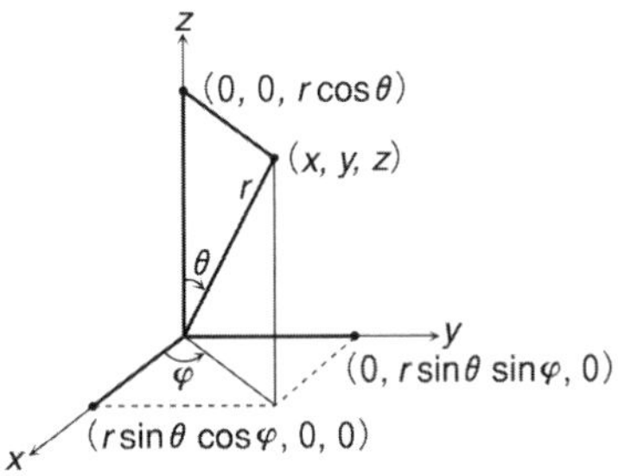

Fig. 3.1. Correspondence between the rectangular and polar coordinates

$$\frac{\partial^2}{\partial x^2}+\frac{\partial^2}{\partial y^2}+\frac{\partial^2}{\partial z^2}=\frac{\partial^2}{\partial r^2}+\frac{2\partial}{r\partial r}-\frac{\hat{\boldsymbol{l}}^2}{r^2\hbar^2}$$

$$\hat{\boldsymbol{l}}^2 \equiv -\hbar^2\left[\frac{1}{\sin\theta}\frac{\partial}{\partial\theta}\left(\sin\theta\frac{\partial}{\partial\theta}\right)+\frac{1}{\sin^2\theta}\frac{\partial^2}{\partial\varphi^2}\right] \tag{3.2}$$

Thus, the equation to be solved becomes:

$$\left[-\frac{\hbar^2}{2m}\left(\frac{\partial^2}{\partial r^2}+\frac{2\partial}{r\partial r}-\frac{\hat{\boldsymbol{l}}^2}{r^2\hbar^2}\right)+V(r)\right]\phi(\boldsymbol{r})=E\phi(\boldsymbol{r}) \tag{3.3}$$

The first step in solving the above equation is to anticipate success in the separation of variables, i.e., it is guessed that $\phi(\boldsymbol{r})$ can be factored out to a radial $R(r)$ and an angular $Y(\theta,\varphi)$ component (cf. Sect. 2.1). When $\phi(\boldsymbol{r}) = R(r)Y(\theta,\varphi)$ is substituted into (3.3), it is apparent that the operation of $\hat{\boldsymbol{l}}^2$ in (3.2) is effective only for the factor of $Y(\theta,\varphi)$. Likewise, the first and second terms of the right-hand side of the first member of (3.2) operate only on the factor of $R(r)$. Thus, (3.3) is rewritten as below.

$$-\frac{\hbar^2}{2m}Y(\theta,\varphi)\left[\frac{\partial^2}{\partial r^2}R(r)+\frac{2\partial}{r\partial r}R(r)\right]+\frac{R(r)}{2mr^2}\hat{\boldsymbol{l}}^2Y(\theta,\varphi)+V(r)R(r)Y(\theta,\varphi)$$
$$=ER(r)Y(\theta,\varphi) \tag{3.4}$$

Dividing both sides by $R(r)Y(\theta,\varphi)$ and rearranging the result leads to the following equation, which shows that the left-hand side is exclusively a function of r whereas the right-hand side depends only on the angles θ and φ.

$$\frac{r^2}{R(r)}\frac{\mathrm{d}^2R(r)}{\mathrm{d}r^2}+\frac{2r}{R(r)}\frac{\mathrm{d}R(r)}{\mathrm{d}r}+\frac{2mr^2}{\hbar^2}[E-V(r)]=\frac{\hat{\boldsymbol{l}}^2Y(\theta,\varphi)}{\hbar^2Y(\theta,\varphi)} \tag{3.5}$$

In order for this equation to be valid, both sides should be constant depending neither on the radius r nor on the angles θ and φ. Furthermore, it is obvious that both sides are dimensionless. If this constant is written as λ, the following two equations are derived:

$$\frac{r^2}{R(r)}\frac{\mathrm{d}^2R(r)}{\mathrm{d}r^2}+\frac{2r}{R(r)}\frac{\mathrm{d}R(r)}{\mathrm{d}r}+\frac{2mr^2}{\hbar^2}[E-V(r)]-\lambda=0 \tag{3.6}$$

$$\hat{\boldsymbol{l}}^2Y(\theta,\varphi)=\lambda\hbar^2Y(\theta,\varphi) \tag{3.7}$$

Therefore, our task becomes the solution of the above two eigenvalue equations. In the subsequent Sects. 3.2.2 and 3.2.3, the equation of the angular component will be discussed first, followed in Sect. 3.2.4, with an examination of the radial component.

3.2.2 An Eigenvalue Equation to Describe the Angular Component (A Digression): Orbital Angular Momentum Operators from the View Point of Classical Mechanics

As mentioned in Sect. 1.4.2, quantum mechanical operators are not derived logically but are arrived at through necessity. In order to familiarize the readers with them I will digress a little in this section prior to the quantum mechanical description in the next section, i.e., I will compare the operator $\hat{\boldsymbol{l}}^2$ to the classical orbital angular momentum.

The left-hand side of the first member of (3.2) can be written as follows by referring to the definition of the momentum operator $\partial/\partial x=\mathrm{i}\hat{p}_x/\hbar$, $\partial/\partial y=\mathrm{i}\hat{p}_y/\hbar$, and $\partial/\partial z=\mathrm{i}\hat{p}_z/\hbar$ (cf. (3.17)).

$$\frac{\partial^2}{\partial x^2}+\frac{\partial^2}{\partial y^2}+\frac{\partial^2}{\partial z^2}=-\frac{\hat{p}^2}{\hbar^2} \tag{3.8}$$

As a trial, we can compare this equation with the classical expression for the kinetic energy of the electron $p^2/(2m)$. For the moment, the electron is regarded as a classical particle and we consider the momentum $\boldsymbol{p}$ at a distance of r from the nucleus (cf. Fig. 1.3). In reference to Fig. 1.3 the momentum is resolved into two components $p_\parallel$ and $p_\perp$. Then, the kinetic energy is written as $p^2/(2m)=p_\parallel^2/(2m)+p_\perp^2/(2m)$. The angular component $p_\perp$ is further rewritten as $p_\perp=mv_\perp=mr\omega$ where the tangential velocity component $v_\perp$ is given as a product of the radius r and the angular velocity $\omega=\mathrm{d}\theta/\mathrm{d}t$. As for the classical orbital angular momentum l_{cls}, it is written from Fig. 1.3 as $l_{cls}=rp_\perp=mr^2\omega$. Thus, the angular component of kinetic energy $p_\perp^2/(2m)$ is equated with $l_{cls}^2/(2mr^2)$, from which the classical kinetic energy of the electron is given as follows.

$$p^2/(2m)=p_\parallel^2/(2m)+p_\perp^2/(2m)=p_\parallel^2/(2m)+l_{cls}^2/(2mr^2) \tag{3.9}$$

This classical kinetic energy is set to correspond to the quantum mechanical kinetic operator $\hat{T}$ in (3.3), i.e.,

$$\hat{T}=-\frac{\hbar^2}{2m}\left(\frac{\partial^2}{\partial r^2}+\frac{2\partial}{r\partial r}\right)+\frac{\hat{\boldsymbol{l}}^2}{2mr^2}\,. \tag{3.10}$$

Comparison of (3.9) and (3.10) suggests that the first term of the right-hand side of (3.10) is an expression of the operator corresponding to the radial component of the classical kinetic energy $p_{\|}^2/(2m)$. In fact, since this first term is equivalent to the operation of $\{\hbar(\partial/\partial r + 1/r)/\mathrm{i}\} \times \{\hbar(\partial/\partial r + 1/r)/\mathrm{i}\}$ multiplied by a constant of $1/(2m)$, the term may be written as $\hat{p}_{\|}^2/(2m)$ where $\hat{p}_{\|} \equiv \{\hbar(\partial/\partial r + 1/r)/\mathrm{i}\}$.

It may thus be recognized that the operator $\hat{\boldsymbol{l}}^2$ appearing in the second term of (3.10) corresponds to the square of the classical orbital angular momentum l_{cls}^2 appearing in (3.9), which is concerned with the energy for the *angular change* in the case of the classical motion.

3.2.3 The Eigenvalue Equation to Describe the Angular Component: Eigenfunctions and Eigenvalues

Returning from the digression in the previous section, we continue the examination of the equations (3.6) and (3.7) in Sect. 3.2.1. First, we start with (3.7) for the orbital angular momentum. Substituting the definition of $\hat{\boldsymbol{l}}^2$ in (3.2) to (3.7), the following eigenvalue equation is obtained:

$$-\hbar^2 \left[\frac{1}{\sin\theta}\frac{\partial}{\partial\theta}\left(\sin\theta\frac{\partial}{\partial\theta}\right) + \frac{1}{\sin^2\theta}\frac{\partial^2}{\partial\varphi^2} \right] Y(\theta,\varphi) = \lambda\hbar^2 Y(\theta,\varphi) \qquad (3.11)$$

This equation shows that the eigenfunction to describe the angular component of the electron is associated with the eigenvalue of $\lambda\hbar^2$. Fortunately, (3.11) is of the same form as the well known equation for describing functions on the surface of spherical objects. The functions $Y(\theta,\varphi)$ are named the spherical surface harmonics, which have been studied in such macroscopic problems as the propagation of heat and waves, and the distribution of electric potential on spherical surfaces. With the use of an infinite series of spherical surface harmonics any spherical function can be described on account of the completeness inherent to the eigenfunctions of the eigenvalue equations discussed in Sect. 2.1.10.

The eigenfunction in (3.11) to describe the angular component of an electron is subject to the boundary conditions such that the function $Y(\theta,\varphi)$ must be finite at $\theta = 0$ and π. From this restriction, it can be mathematically derived that the dimensionless λ in (3.11) must be equal to $\lambda = l(l+1)$ with $l = 0, 1, 2, \ldots$, although the derivation will be omitted here. Furthermore, the value of l can be shown to terminate at a certain integer $n-1$ when the potential $V(r)$ in (3.1) or (3.3) is Coulombic. Again, the verification of this conclusion will be omitted, but related arguments will be made in Sects. 3.3.4 and 3.4.2, and the physical meaning of the above integer n will be explained in Sect. 3.2.4.

In continuation, in the solution of the eigenvalue equation (3.11), the eigenfunction $Y(\theta,\varphi)$ is supposed to be separable as $Y(\theta,\varphi) = \Theta(\theta)\Phi(\varphi)$, which turns out to be the case. Applying the method of separation of variables in Sect. 3.2.1, the problem of solving (3.11) is now reduced to solving the following two eigenvalue equations, where the symbol m^2 is a dimensionless constant arising from the separation.

$$\left[\sin\theta\frac{\mathrm{d}}{\mathrm{d}\theta}\left(\sin\theta\frac{\mathrm{d}}{\mathrm{d}\theta}\right)+l(l+1)\sin^2\theta\right]\Theta(\theta)=m^2\Theta(\theta) \tag{3.12}$$

$$\frac{\mathrm{d}^2\Phi(\varphi)}{\mathrm{d}\varphi^2}=-m^2\Phi(\varphi) \tag{3.13}$$

Needless to say, the above symbol m has nothing to do with the mass m introduced in the note at the end of Sect. 3.1. The symbol here is, in fact, the same as that appearing in Fig. 1.4, where the symbol was used to differentiate the physically observable split state in a magnetic field. For this reason the symbol is named the magnetic quantum number.

As for the other separation constant λ in (3.6), it was already related to another quantum number l by $\lambda = l(l+1)$. This quantum number differentiates the directional difference in the position of the electron, as will be demonstrated in Sects. 3.3.2 and 3.3.3. For this reason it is called the azimuthal quantum number, or simply, the subordinate quantum number.

Equation (3.13) is immediately solved to give $\Phi(\varphi) = \text{const}\times \mathrm{e}^{-im\varphi}$. Since $\Phi(\varphi)$ must be single-valued as stated in Sect. 1.2, the value of m is restricted to $m = 0, \pm1, \pm2, \ldots$; if m is not an integer, $\Phi(\varphi)$ cannot be single-valued when the variable changes from φ to, e.g., $\varphi + 2\pi$. On the other hand, the value of m cannot extend to infinity but is limited at $\pm l$ by the boundary conditions that the spherical surface harmonics should not diverge at $\theta = 0$ and π. The same requirement can be derived from the formal discussion of the angular momentum in general (cf. Sect. 3.4).

All in all, to specify the spherical surface harmonics in (3.11) the two quantum numbers l and m must be given. The factor $\Theta(\theta)$ in $Y(\theta,\varphi)$ is equal to what is called the associated Legendre polynomial, which has long been known and is designated usually as $P_l^{|m|}(\cos\theta)$. Thus, the spherical surface harmonics can be written as $Y_{l,m}(\theta,\varphi) = \text{const} \times P_l^{|m|}(\cos\theta)\mathrm{e}^{\mathrm{i}m\varphi}$. The constant is for the normalization explained in Sect. 2.1.5, in general. In this way the eigenfunction $Y_{l,m}(\theta,\varphi)$ as well as the eigenvalue $\lambda\hbar^2$ with $\lambda = l(l+1)$ and $l = 0, 1, 2, 3, \ldots, (n-1)$ are fixed rigidly for the operator $\hat{\boldsymbol{l}}^2$.

Let us move to another operator related to the orbital angular momentum. As stated in connection with Fig. 1.3 in Chap. 1, the orbital angular momentum $\boldsymbol{l}$ is expressed as $\boldsymbol{l} = l_x\boldsymbol{i} + l_y\boldsymbol{j} + l_z\boldsymbol{k}$ in the rectangular coordinate system. Accordingly, the operator $\hat{\boldsymbol{l}}$ is also written as $\hat{\boldsymbol{l}} = \hat{l}_x\boldsymbol{i} + \hat{l}_y\boldsymbol{j} + \hat{l}_z\boldsymbol{k}$ where $\hat{l}_z$, e.g., is defined as $\hat{l}_z = x\hat{p}_y - y\hat{p}_x$. On account of the unique correspondence between the rectangular and the polar coordinate systems, the operator $\hat{l}_z$ can be replaced with an expression in the polar coordinates (after tedious replacement of partial differentials, which is omitted here) to obtain the result $\hat{l}_z = -\mathrm{i}\hbar\partial/\partial\varphi$. When this operator operates on $Y_{l,m}(\theta,\varphi) = \text{const} \times P_l^{|m|}(\cos\theta)\mathrm{e}^{\mathrm{i}m\varphi}$, the following result is obtained:

$$\hat{l}_z Y_{l,m}(\theta,\varphi) = \frac{\hbar}{\mathrm{i}}\frac{\partial}{\partial\varphi}Y_{l,m}(\theta,\varphi) = m\hbar Y_{l,m}(\theta,\varphi) \tag{3.14}$$

This equation indicates that the function $Y_{l,m}(\theta,\varphi)$ is not only an eigenfunction of the energy but also a simultaneous eigenfunction for both operators $\hat{\boldsymbol{l}}^2$ and $\hat{l}_z$, with the respective eigenvalues of $l(l{+}1)\hbar^2$, $l{=}0,1,2,3,\ldots,(n{-}1)$ and $m\hbar$, $m = 0, \pm1, \pm2, \ldots, \pm l$. As a result, when the system under consideration is put into a magnetic field with a magnetic flux density of $\boldsymbol{B} = (0, 0, B)$, the energy of the system is split into $2l{+}1$ states according to the relationship $E = -\mu_z B = -\gamma m\hbar B$. Note, however, the electron spin has been set aside for the moment.

Note here that the operator $\hat{l}_z$ above could have been chosen as $\hat{l}_x$ or $\hat{l}_y$ because in transforming the rectangular coordinates to the polar coordinates the choice of one of the three axes x, y, z is arbitrary. The essential point is that the function $Y_{l,m}(\theta,\varphi)$ is the simultaneous eigenfunction of $\hat{\boldsymbol{l}}^2$ and one of the three components of the vectorial operator $\hat{\boldsymbol{l}}$.

In this way the angular component of the eigenfunction $\phi(\boldsymbol{r})$ in (3.3) is completely solved. The explicit form of normalized functions $Y_{l,m}(\theta,\varphi)$ is listed in Table 3.1 to be given in the subsequent Sect. 3.2.5. The first few members are preliminarily shown in (3.15). The sign of the subscript for $Y_{1,\pm1}$ has no physical meaning. However, since there is an arbitrariness in the phase convention of the spherical surface harmonics, the Condon-Shortley (cf. Refs. for Chap. 3) convention is adopted as in the last member of (3.15). With this choice of convention the consistency of sign is conserved leading unambiguously to such relations as in the set of (3.20) through (3.21) and in those of (3.29) and (3.30).

$$\begin{aligned} &Y_{0,0} = 1/(2\sqrt{\pi}), \quad Y_{1,0} = \sqrt{3}\cos\theta/(2\sqrt{\pi}), \\ &Y_{1,\pm1} = \mp\sqrt{3}\sin\theta\cdot\exp(\pm\mathrm{i}\varphi)/(2\sqrt{2\pi}), \ldots, \\ &Y_{l,-m}(\theta,\varphi) = (-1)^m Y_{l,m}^*(\theta,\varphi) \end{aligned} \tag{3.15}$$

3.2.4 The Eigenvalue Equation to Describe the Radial Component: Eigenfunctions and Eigenvalues

Since the separation constant λ in (3.6) is now fixed as $l(l+1)$, a concrete form of the eigenvalue equation for the radial component $R(r)$ can be derived from (3.6) by replacing λ with $l(l+1)$ and by substituting $V(r)$ with the explicit form of $-Ze^2/(4\pi\varepsilon_0 r)$. Fortunately, this eigenvalue equation also can be solved analytically. The procedure to solve the equation will be briefly explained in the following.

As a mathematical technique the variable r in (3.6) having the dimension of length is replaced with a dimensionless variable $\rho = \alpha r$ where

$\alpha \equiv \sqrt{8m|E|}/\hbar$ is a constant having the reciprocal dimension of length. Similarly, in order to make the Coulombic potential term also dimensionless, a dimensionless parameter β defined as $\beta \equiv \sqrt{m/(2|E|)}e^2/(4\pi\varepsilon_0\hbar)$ is introduced. Then, (3.6) is converted to (3.16) below which is an equation in terms of dimensionless quantities ρ and β. The sought function $R(r)$ in (3.6) is now equated to $S(\rho)$ with the restriction of $0 \leq \rho \leq \infty$.

$$\frac{1}{\rho^2}\frac{\mathrm{d}}{\mathrm{d}\rho}\left(\rho^2\frac{\mathrm{d}S(\rho)}{\mathrm{d}\rho}\right) + \left[\frac{\beta}{\rho} - \frac{1}{4} - \frac{l(l+1)}{\rho^2}\right]S(\rho) = 0 \tag{3.16}$$

Further mathematical manipulation to solve $S(\rho)$ requires the replacement of $S(\rho)$ with $S(\rho) = \rho^l \mathrm{e}^{-\rho/2}L(\rho)$, and the function $L(\rho)$ is then sought. In this process, it is required that $S(\rho)$ must not diverge at the origin $\rho = 0$, and must converge when $\rho = \infty$, which are boundary conditions that should be familiar by now. As a result of the requirement, it is derived automatically that the values of β must be positive integers $1, 2, 3, \ldots$. When this integer is written as n, the relation $\beta = n = \sqrt{m/(2|E|)}e^2/(4\pi\varepsilon_0\hbar)$ yields the following equation for the eigenvalue of energy:

$$E = -\frac{Z^2me^4}{2(4\pi\varepsilon_0\hbar)^2n^2} \tag{3.17}$$

This result is in complete agreement with the result derived by Bohr in the old quantum mechanical treatment as described in Sects. 1.4 and 3.1. Note that the letter m_e in (3.11) is replaced with m according to the statement in the note at the end of Sect. 3.1. Equation (3.17) above shows that the energy E is determined exclusively by the positive integer n with no dependence on the azimuthal quantum number l, which is introduced in Sect. 3.2.3. The integer n is called the principal quantum number (or the total quantum number).

The eigenfunction $R(r)$ corresponding to the eigenvalue E is characterized by the function $L(\rho)$ as shown above. This function also has long been known as an associated Laguerre polynomial. Since the eigenfunction $R(r)$ is supposed to depend on the quantum numbers n and l through the definitions of β and $S(\rho)$, the eigenfunction is denoted as $R_{n,l}(r)$. The third column of Table 3.1 lists the initial several normalized eigenfunctions explicitly. All of them are expressed as a product of polynomials of r with an exponential function of r. In (3.18) the first few radial functions are extracted from Table 3.1. The symbol a_0 stands for the Bohr radius, which is equal to $4\pi\varepsilon_0\hbar^2/(m_\mathrm{e}e^2)$. With the use of the Bohr radius the proportionality constant α in $\rho = \alpha r$ is written as $\alpha = 2Z/(na_0)$.

$$\begin{aligned} R_{1,0} &= 2\exp\{-(Zr/a_0)\} \times (Z/a_0)^{3/2} \\ R_{2,0} &= (1/2\sqrt{2})\{2 - (Zr/a_0)\}\exp\{-(Zr/a_0)/2\} \times (Z/a_0)^{3/2} \\ R_{2,1} &= (1/2\sqrt{6})(Zr/a_0)\exp\{-(Zr/a_0)/2\} \times (Z/a_0)^{3/2} \end{aligned} \tag{3.18}$$

3.2.5 Total Spatial Eigenfunction of the One-Electron Atom

The total spatial eigenfunction $\phi(\boldsymbol{r})$ of one-electron atoms is given by the product of the eigenfunction $Y_{l,m}(\theta, \varphi)$ for the angular component and the eigenfunction $R_{n,l}(r)$ for the radial component. Table 3.1 shows the two components explicitly up to $n = 3$. The dimension of $\phi(\boldsymbol{r}) = R_{n,l}(r)Y_{l,m}(\theta, \varphi)$ is equal to the dimension of the reciprocal of square root of volume as stated in Sect. 2.1.1, but since $Y_{l,m}(\theta, \varphi)$ is dimensionless, the dimension of $\phi(\boldsymbol{r})$ is determined by the factor $\times(Z/a_0)^{3/2}$ in $R_{n,l}(r)$. It is seen from Table 3.1 that some of the spherical surface harmonics $Y_{l,m}(\theta, \varphi)$ are complex numbers. These will be explained later in Sect. 3.3.2. At this moment only those in the real domain will be examined.

According to Table 3.1, only the case $l = m = 0$ is possible for the state $n = 1$. Therefore, the total eigenfunction is limited exclusively to $\phi(\boldsymbol{r}) = R_{1,0}(r)Y_{0,0}$, which is an isotropic function of the radius r only.

As shown in Fig. 3.2(a), the eigenfunction is a simple exponential function of r. In Fig. 3.2(a), the z-axis of the rectangular xyz-coordinate is appropriated to show the magnitude of $R_{1,0}(r)$ and the xy-plane is utilized to project the magnitude as a contour map. The angular component $Y_{0,0}$ is isotropic with a radius of $1/(2\sqrt{\pi})$.

When $n = 2$, l is either 0 or 1. When $l = 0$, only the case $m = 0$ is possible so that again a sphere of $\phi(\boldsymbol{r}) = R_{2,0}(r)Y_{0,0}$ is obtained, but in this case $R_{2,0}(r)$ becomes 0 at $r = 2a_0/Z$. Accordingly, the total eigenfunction $\phi(\boldsymbol{r})$ changes its sign once along the direction of the radius. The spherical plane, across which there is a sign change of the eigenfunction, is generally called the nodal plane.

In the case of $n = 2$, $l = 1$, there are three states, with $m = 0, \pm 1$. The discussion for the case $m = -1$ is postponed until Sect. 3.3.2 because the eigenfunctions become complex numbers. Only the real form of $\phi(\boldsymbol{r}) = R_{2,1}(r)Y_{1,0}(0, \varphi)$ is explained below.

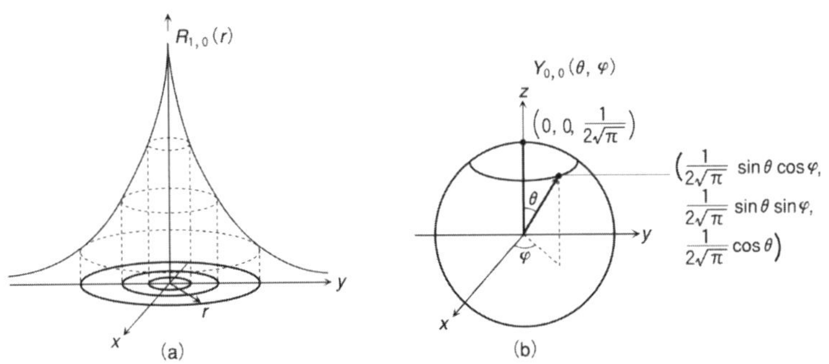

Fig. 3.2. (**a**) Illustration of the radial and (**b**) the angular components of the total eigenfunction $\phi(\boldsymbol{r}) = R_{n,l}(r)Y_{l,m}(\theta, \varphi)$ for the state of $n = 1$ of one-electron atoms

Table 3.1. Radial ($R_{n,l}(r)$) and angular ($Y_{l,m}(\theta,\varphi)$) eigenfunctions of one-electron atoms of the atomic number Z

n	l	m	Symbol	$R_{n,l}(r)$ ($a_0 = 4\pi\varepsilon_0\hbar^2/(m_e e^2)$)	$Y_{l,m}(\theta,\varphi)$
1	0	0	1s	$2\exp\left\{-\left(\frac{Zr}{a_0}\right)\right\}\times\left(\frac{Z}{a_0}\right)^{3/2}$	$\frac{1}{2\sqrt{\pi}}$
2	0	0	2s	$\frac{1}{2\sqrt{2}}\left\{2-\left(\frac{Zr}{a_0}\right)\right\}\exp\left\{-\left(\frac{Zr}{a_0}\right)/2\right\}\times\left(\frac{Z}{a_0}\right)^{3/2}$	$\frac{1}{2\sqrt{\pi}}$
2	1	0	2p$_0$	$\frac{1}{2\sqrt{6}}\left(\frac{Zr}{a_0}\right)\exp\left\{-\left(\frac{Zr}{a_0}\right)/2\right\}\times\left(\frac{Z}{a_0}\right)^{3/2}$	$\frac{\sqrt{3}}{2\sqrt{\pi}}\cos\theta$
2	1	±1	2p$_{\pm1}$	$\frac{1}{2\sqrt{6}}\left(\frac{Zr}{a_0}\right)\exp\left\{-\left(\frac{Zr}{a_0}\right)/2\right\}\times\left(\frac{Z}{a_0}\right)^{3/2}$	$\mp\frac{\sqrt{3}}{2\sqrt{2\pi}}\sin\theta\cdot\exp(\pm i\varphi)$
3	0	0	3s	$\frac{2}{81\sqrt{3}}\left\{27-18\left(\frac{Zr}{a_0}\right)+2\left(\frac{Zr}{a_0}\right)^2\right\}\times\exp\left\{-\left(\frac{Zr}{a_0}\right)/3\right\}\times\left(\frac{Z}{a_0}\right)^{3/2}$	$\frac{1}{2\sqrt{\pi}}$
3	1	0	3p$_0$	$\frac{2\sqrt{2}}{81\sqrt{3}}\left\{6-\left(\frac{Zr}{a_0}\right)\right\}\left(\frac{Zr}{a_0}\right)\times\exp\left\{-\left(\frac{Zr}{a_0}\right)/3\right\}\times\left(\frac{Z}{a_0}\right)^{3/2}$	$\frac{\sqrt{3}}{2\sqrt{\pi}}\cos\theta$
3	1	±1	3p$_{\pm1}$	$\frac{2\sqrt{2}}{81\sqrt{3}}\left\{6-\left(\frac{Zr}{a_0}\right)\right\}\left(\frac{Zr}{a_0}\right)\times\exp\left\{-\left(\frac{Zr}{a_0}\right)/3\right\}\times\left(\frac{Z}{a_0}\right)^{3/2}$	$\mp\frac{\sqrt{3}}{2\sqrt{2\pi}}\sin\theta\cdot\exp(\pm i\varphi)$
3	2	0	3d$_0$	$\frac{2\sqrt{2}}{81\sqrt{15}}\left(\frac{Zr}{a_0}\right)^2\exp\left\{-\left(\frac{Zr}{a_0}\right)/3\right\}\times\left(\frac{Z}{a_0}\right)^{3/2}$	$\frac{\sqrt{5}}{4\sqrt{\pi}}(3\cos^2\theta-1)$
3	2	±1	3d$_{\pm1}$	$\frac{2\sqrt{2}}{81\sqrt{15}}\left(\frac{Zr}{a_0}\right)^2\exp\left\{-\left(\frac{Zr}{a_0}\right)/3\right\}\times\left(\frac{Z}{a_0}\right)^{3/2}$	$\mp\frac{\sqrt{15}}{2\sqrt{2\pi}}\sin\theta\cos\theta\cdot\exp(\pm i\varphi)$
3	2	±2	3d$_{\pm2}$	$\frac{2\sqrt{2}}{81\sqrt{15}}\left(\frac{Zr}{a_0}\right)^2\exp\left\{-\left(\frac{Zr}{a_0}\right)/3\right\}\times\left(\frac{Z}{a_0}\right)^{3/2}$	$\frac{\sqrt{15}}{4\sqrt{2\pi}}\sin^2\theta\cdot\exp(\pm i2\varphi)$

Table 3.1. (condinued.)

Note: The total eigenfunction $\phi(\boldsymbol{r})$ is given by the product of the two normalized components $R_{n,l}(r)$ and $Y_{l,m}(\theta,\varphi)$. The symbols s, p, d, ... appearing in the second column of Table 3.1 as 1s, 2p, 3d, ... correspond, respectively, to the azimuthal quantum numbers of $l = 0, 1, 2, \ldots$. Historically, the notations were introduced by spectroscopists. If the effects due to the relativisty and to the electron and proton spins which cause very minute splittings of energy level are ignored, there is no energy splitting in one-electron atoms because of the independence of energy on the azimuthal quantum number l. In multi-electron atoms, however, the interaction among the electrons causes energy splittings which vary with l, as will be shown in Chap. 4. For example, in the absorption spectrum of the Li atom (cf. Fig. 4.2), the transitions from a state corresponding to $n = 2$ and $l = 0$ of the H atom to states of $n \geq 2$ and $\underline{l = 1}$ is conspicuous so that the lines due to these transitions were named principal series. Hence, states with $l = 1$ are designated with the letter 'p'. Likewise, since the sharp lines associated with the transitions from $n = 2$ and $l = 1$ to $n \geq 3$ and $\underline{l = 0}$ were called sharp series, states with $l = 0$ are represented with the letter 's'. Furthermore, states with $l = 2$ are denoted by 'd', because of the diffuse lines corresponding to the transitions from $n = 2$ and $l = 1$ to $n \geq 3$ and $\underline{l = 2}$.

Figure 3.3(a) demonstrates that when two spheres of diameter $\sqrt{3}/(2\sqrt{\pi})$ are placed above and below the xy-plane along the z-axis, a vector originating from the origin and having a length of $\sqrt{3}\cos\theta/(2\sqrt{\pi})$ depicts two circles at the intersections of the spheres and the cones. Figure 3.3(b) shows approximately the function $R_{2,1}(r)$ in Table 3.1 as a function of radius r parallel with the vector in Fig. 3.3(a).

Since the total eigenfunction $\phi(\boldsymbol{r})$ is given by the product of two components shown in Figs. 3.3(a) and (b), it can be represented as in Fig. 3.3(c)

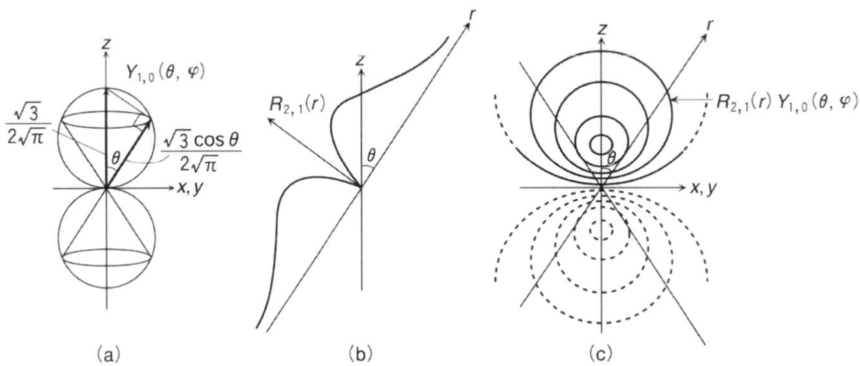

Fig. 3.3. (**a**) Illustration of the angular and (**b**) the radial components of the total eigenfunction $\phi(\boldsymbol{r}) = R_{n,l}(r)Y_{l,m}(\theta,\varphi)$ for the state of $n = 2$, $l = 1$ and $m = 0$ of one-electron atoms. (**c**) The product of the two components

as a contour map, where the solid and broken lines indicate that $\phi(\boldsymbol{r})$ is positive above the xy-plane $(0 \leq \theta < \pi/2)$ and negative below the plane $(\pi/2 < \theta \leq \pi)$ according to the sign of $\cos\theta$.

As shown in the above examples, spherical surface harmonics in (3.15) can be visually demonstrated when they are in the real form, but not in the complex form. However, as stated in Sect. 1.2 and Chap. 2, the physical meaning is given to the eigenfunction only when the square of the absolute value is dealt with. In other words, it is not necessary to worry about the fact that eigenfunctions in the imaginary form can not be depicted in a visible way. Some topics related to this statement will be discussed in Sects. 3.3.2 and 3.3.3.

3.3 Some Comments on the Solution for One-Electron Atoms

3.3.1 Degeneracy of the Spatial Eigenfunctions is n^2-Fold without the Electron Spin Components, and is $2n^2$-Fold with the Electron Spin Included

So far it has been shown that the spatial eigenfunction $\phi(\boldsymbol{r}) = R_{n,l}(r)Y_{l,m}(\theta,\varphi)$ is characterized by the three quantum numbers n, l, m, while the eigenvalue E given by (3.17) depends only on n. The other two quantum numbers are limited as $l = 0, 1, 2, \ldots, (n-1)$ and $m = 0, \pm 1, \pm 2, \ldots, \pm l$ as explained in Sect. 3.2.3. Thus, the total number of eigenfunctions associated with an energy of E_n is given as n^2 by (3.19) below.

$$\sum_{l=0}^{n-1}(2l+1) = n^2 \tag{3.19}$$

In other words, the degeneracy of the spatial component is n^2-fold. This degeneracy is proper to the eigenfunction for a Coulombic potential, and is called accidental degeneracy* (V.A. Fock, 1935). In multi-electron atoms, to be discussed in Chap. 4, the potential is non-Coulombic, and the accidental degeneracy is removed to give different energy values for states of the same n but different l values.

As is mentioned in Sects. 1.3 and 3.4, the number of quantum number values used to characterize the spin state of an electron is just 2, and the states are energetically indistinguishable in the absence of an external magnetic field. Thus, the degeneracy of the overall eigenfunction including the spin component is $2n^2$-fold. The above statement, however, is applicable only to cases where the spin-orbit interaction to be discussed in Sect. 4.3 is negligibly small.

3.3.2 Representation of Real Forms of Spherical Surface Harmonics

As (3.15) indicates, the spherical surface harmonics to describe the angular components of eigenfunctions are, in general, complex numbers, which is anticipated from the fact that the operator $\hat{l}_z$ involves the imaginary symbol i. Spherical surface harmonics in complex numbers, however, can be converted to real numbers without losing their physical significance to describe the angular component of eigenfunctions; for each value of l of $Y_{l,m}(\theta,\varphi)$, there is a pair of $\pm m$, which are energetically degenerate in the absence of magnetic fields. According to Sect. 2.1.7, any linear combination of these degenerate functions is also associated with the same degenerate energy as that of the original pair. Thus, in reference to (3.15) and the Euler equations, $\cos m\varphi = (\mathrm{e}^{\mathrm{i}m\varphi}+\mathrm{e}^{-\mathrm{i}m\varphi})/2$ and $\sin m\varphi = (\mathrm{e}^{\mathrm{i}m\varphi}-\mathrm{e}^{-\mathrm{i}m\varphi})/2\mathrm{i}$, the original pair $Y_{l,m}$ and $Y_{l,-m}$ can be replaced with a pair of real form, which describes the same energy state as the original pair.

An example case of $l = 1$ is given, after normalization, as follows:

$$-Y_{1,1}/\sqrt{2}+Y_{1,-1}/\sqrt{2} = \sqrt{3}\sin\theta\cos\varphi/(2\sqrt{\pi}) \equiv Y_{1x} \tag{3.20}$$

$$\mathrm{i}Y_{1,1}/\sqrt{2}+\mathrm{i}Y_{1,-1}/\sqrt{2} = \sqrt{3}\sin\theta\sin\varphi/(2\sqrt{\pi}) \equiv Y_{1y} \tag{3.21}$$

When the above two real forms are compared with the real form $Y_{1,0} = \sqrt{3}\cos\theta/(2\sqrt{\pi}) \equiv Y_{1z}$, shown in Fig. 3.3(a), it is seen that all three real forms express the length of the vector from the origin to two spheres arranged like a dumbbell parallel to the x, y, z axes, respectively. This situation is analogous to the geometrical relations of $x = r\sin\theta\cos\varphi$, $y = r\sin\theta\sin\varphi$, and $z = r\cos\theta$ in Fig. 3.1. Therefore, from the functions designated as $\mathrm{p}_0, \mathrm{p}_{\pm1}, \mathrm{d}_0, \mathrm{d}_{\pm1}$, and $\mathrm{d}_{\pm2}$ in Table 3.1, the functions in the real form are derived as in Table 3.2.

By multiplying the radial component $R_{n,l}(r)$ in Table 3.1 with the spherical surface harmonics in the real form given in Table 3.2, the total spatial eigenfunctions $\phi(\boldsymbol{r})$ are obtained, which are illustrated in Figs. 3.4 and 3.5.

In both figures the gray and the white surfaces on the encapsulating spheres are drawn to facilitate recognition of the sign change in the lobes of the p- and d-functions enclosed by the spheres. The solid and the broken lines correspond to lobes with positive and negative signs. From the two figures it is seen that the total eigenfunctions become null $l-|m|$ times as the

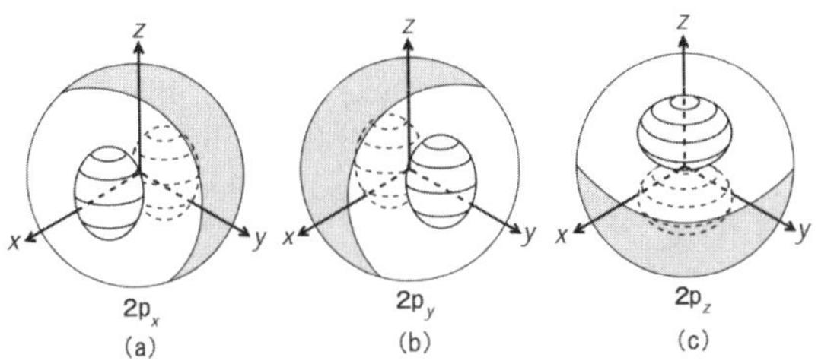

Fig. 3.4. An illustration of the p-functions in the real form

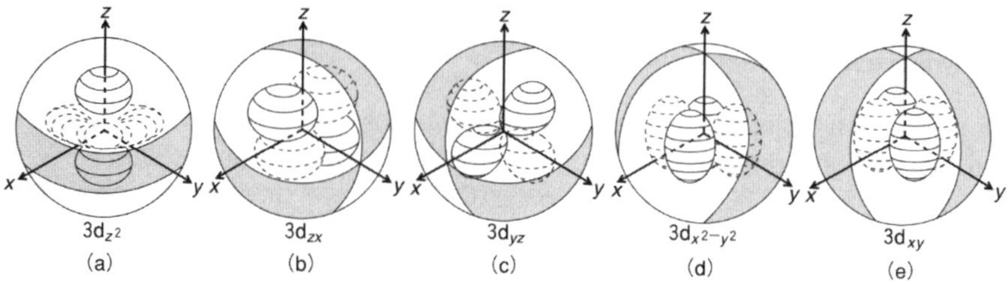

Fig. 3.5. An illustration of the d-functions in the real form

Table 3.2. Expressions of the spherical surface harmonics in the real form

$Y_{1,0} = \frac{\sqrt{3}}{2\sqrt{\pi}} \cos\theta = \frac{\sqrt{3}}{2\sqrt{\pi}} \frac{z}{r}$	p_z
$\frac{-Y_{1,1}}{\sqrt{2}} + \frac{Y_{1,-1}}{\sqrt{2}} = \frac{\sqrt{3}}{2\sqrt{\pi}} \sin\theta\cos\varphi = \frac{\sqrt{3}}{2\sqrt{\pi}} \frac{x}{r}$	p_x
$\frac{\mathrm{i}Y_{1,1}}{\sqrt{2}} + \frac{\mathrm{i}Y_{1,-1}}{\sqrt{2}} = \frac{\sqrt{3}}{2\sqrt{\pi}} \sin\theta\sin\varphi = \frac{\sqrt{3}}{2\sqrt{\pi}} \frac{y}{r}$	p_y
$Y_{2,0} = \frac{\sqrt{5}}{4\sqrt{\pi}}(3\cos^2\theta - 1) = \frac{\sqrt{5}}{4\sqrt{\pi}}\left(\frac{3z^2}{r^2} - 1\right)$	d_{z^2}
$\frac{-Y_{2,1}}{\sqrt{2}} + \frac{Y_{2,-1}}{\sqrt{2}} = \frac{\sqrt{15}}{4\sqrt{\pi}} \sin 2\theta\cos\varphi = \frac{\sqrt{15}}{4\sqrt{\pi}} \frac{2zx}{r^2}$	d_{zx}
$\frac{\mathrm{i}Y_{2,1}}{\sqrt{2}} + \frac{\mathrm{i}Y_{2,-1}}{\sqrt{2}} = \frac{\sqrt{15}}{4\sqrt{\pi}} \sin 2\theta\sin\varphi = \frac{\sqrt{15}}{4\sqrt{\pi}} \frac{2yz}{r^2}$	d_{yz}
$\frac{Y_{2,2}}{\sqrt{2}} + \frac{Y_{2,-2}}{\sqrt{2}} = \frac{\sqrt{15}}{4\sqrt{\pi}} \sin^2\theta\cos 2\varphi = \frac{\sqrt{15}}{4\sqrt{\pi}} \frac{x^2 - y^2}{r^2}$	$\mathrm{d}_{x^2-y^2}$
$\frac{Y_{2,2}}{\sqrt{2}\mathrm{i}} + \frac{Y_{2,-2}}{\sqrt{2}\mathrm{i}} = \frac{\sqrt{15}}{4\sqrt{\pi}} \sin^2\theta\sin 2\varphi = \frac{\sqrt{15}}{4\sqrt{\pi}} \frac{2xy}{r^2}$	d_{xy}

polar angle θ varies between 0 and π, and $|m|$ times as the azimuthal angle φ changes between 0 and 2π. Thus, in all, the total eigenfunctions nullify l times.

3.3.3 A Relation between the Angular Dependence of Spherical Harmonics and the Orbital Angular Momentum

While (3.14) demonstrates that the z-axis component of the orbital angular momentum in the states $Y_{1,1}$ and $Y_{1,-1}$ assumes the eigenvalues of $\pm\hbar$, the square of the absolute values of $|Y_{1,1}|^2$ and $|Y_{1,-1}|^2$ in Figs. 3.6(a) and (c) may be regarded as representing the probability of finding the electron which is circulating around the z-axis.

As shown in Fig. 1.3, when a classical particle circulates around the z-axis, the orbital angular momentum vector can change continuously. In contrast,

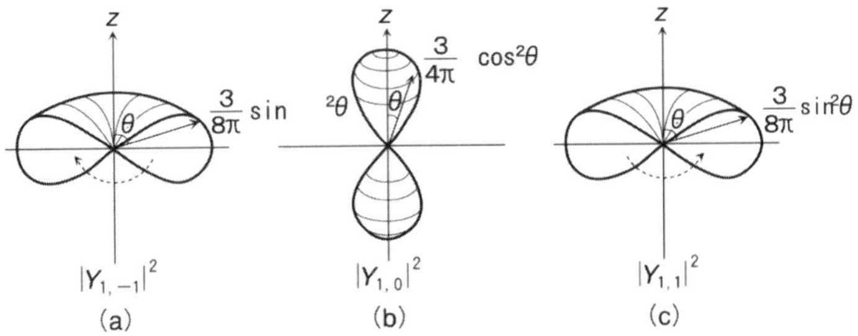

Fig. 3.6. Probability of finding the electron in the state when the z-axis component of orbital angular momentum is $-\hbar$ in (**a**), 0 in (**b**), and $+\hbar$ in (**c**)

when an electron is in the quantum mechanical state described by the eigenfunction $Y_{l,m}$ satisfying (3.11) and (3.14), the orbital angular momentum $\hat{l}$ must be determined by the following eigenvalue equations, which means that only discrete values of $\sqrt{l(l+1)}\hbar$ with its z-axis component of $m\hbar$ are allowed.

$$\hat{l}^2 Y_{l,m} = l(l+1)\hbar^2 Y_{l,m}, \quad \hat{l}_z Y_{l,m} = m\hbar \tag{3.22}$$

Figure 3.7 demonstrates the three states allowed by (3.22) for the case of $l = 1$. While Fig. 3.6 is for the positional space, Fig. 3.7 shows the orbital angular momentum vector space, and values of both $\sqrt{l^2}$ and l_z are shown.

Figures 3.4(a) and (b) may be interpreted as the superposition of the two circulating wave states $Y_{1,1}$ and $Y_{1,-1}$ according to (3.20) and (3.21), which gives two standing waves fixed along the x- and y-axes. Since these standing waves are obtained by linear combination of two degenerate circulating wave states, the energy state of the electron with $l = 1$ can be equally described either by the sets Y_{1x}, Y_{1y} or $Y_{1,1}, Y_{1,-1}$. The choice depends upon the problem under consideration. For the problem of molecules with a fixed geometrical arrangement, e.g., it is more natural to use the set in the real domain.

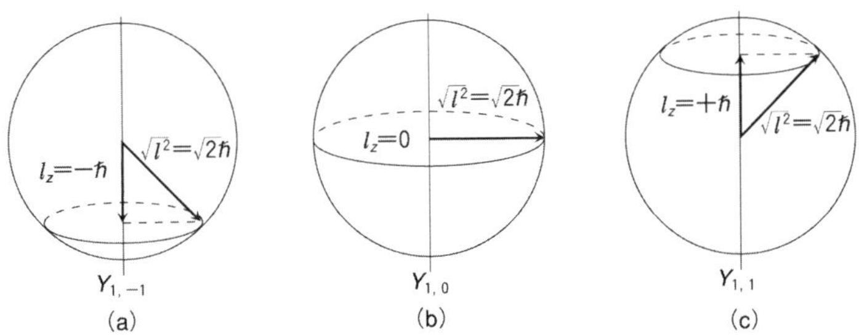

Fig. 3.7. The three states of $l = 1$ represented in the space of the orbital angular momentum vector

As for the orbital angular momentum, however, the set Y_{1x} and Y_{1y} do not qualify as eigenfunctions of $\hat{l}_z$ because the operation of $\hat{l}_z$ in (3.14) on Y_{1x} and Y_{1y} in (3.20) and (3.21) leads to the relationships $\hat{l}_z Y_{1x} = \mathrm{i}Y_{1y}$ and $\hat{l}_z Y_{1y} = -\mathrm{i}Y_{1x}$, respectively, which do not have the form of an eigenvalue equation. Furthermore, these equations show that the expectation value of $\hat{l}_z$ for both Y_{1x} and Y_{1y} is 0 (cf. Sect. 2.1.11 and the orthogonality of Y_{1x} and Y_{1y}). In quantum mechanics such a situation is called the 'quenching of the orbital angular momentum'. The result is similar to the situation where the superposition of the two circulating waves in Fig. 3.6 yields the two standing waves described by Y_{1x} and Y_{1y}. This 'quenching' is another example in support of the statement in Sect. 2.1.12, i.e., not all the eigenfunctions of the commutative operators of $\hat{H}$ and $\hat{l}_z$ coincide in a one-to-one manner.

From Eq. (3.15) the following relations are obtained: $|Y_{0,0}|^2 = 1/(4\pi)$, $|Y_{1,0}|^2 = 3\cos^2\theta/(4\pi)$, $|Y_{1,\pm 1}|^2 = 3\sin^2\theta/(8\pi)$, ... If all $2l+1$ degenerate components belonging to the same value of l are summed up, the result is as follows: $|Y_{1,0}|^2 + |Y_{1,1}|^2 + |Y_{1,-1}|^2 = 3/(4\pi)$, $|Y_{2,0}|^2 + |Y_{2,1}|^2 + |Y_{2,-1}|^2 + |Y_{2,2}|^2 + |Y_{2,-2}|^2 = 5/(4\pi), \ldots, (2l+1)/(4\pi)$ which expresses the completion of an isotropic sphere whose size increases with l. This relation is called Unsöld's theorem* (cf. Sect. 4.2).

3.3.4 The Radial Potential

The eigenvalue equation for the radial component considered in Sects. 3.2.1 and 3.2.4 can be rewritten as follows.

$$-\frac{\hbar^2}{2mr^2}\left\{2r\frac{\mathrm{d}R}{\mathrm{d}r} + r^2\frac{\mathrm{d}^2R}{\mathrm{d}r^2}\right\} + \left\{\frac{l(l+1)\hbar^2}{2mr^2} - \frac{e^2}{4\pi\varepsilon_0 r}\right\}R = E_n R \qquad (3.23)$$

According to the discussion in connection with (3.10), it is recognized that the first term of the left-hand side of (3.23) corresponds to the kinetic energy associated with the radial motion while the first term in the second curly brackets corresponds to the kinetic energy for the angular motion and the rest of the expression in the brackets represents the Coulombic potential energy. The angular kinetic energy can be regarded as a potential energy which is related to the centrifugal force in classical dynamics as follows: the centrifugal force F_c in classical dynamics may be written as $F_\mathrm{c} = mv_\perp^2/r = mr\omega^2 = l_{cls}^2/(mr^3)$ with the use of the same notations as in (3.9). Denoting the potential to yield this force by V_c, and referring to the relation $F_\mathrm{c} = -\mathrm{d}V_\mathrm{c}/\mathrm{d}r$, the potential V_c can be written as $V_\mathrm{c} = l_{cls}^2/(2mr^2)$, which corresponds to the above stated angular kinetic energy. It may be interpreted as an expression of the fact that the electron with this kinetic energy for $l \neq 0$ tends to move away from the nucleus on account of the centrifugal force. For this reason the first term in the second brackets of (3.23) is called the centrifugal potential and the sum in the brackets represents the effective potential for the radial motion of the electron.

The solid lines between the two broken curves in Figs. 3.8(b) and (c) illustrate these effective potentials which are designated as $V_{\rm eff}(r)$. From the figures it is seen that the eigenfunctions $R(r)$ shift gradually outward as the value of l increases as $0, 1, 2, \ldots$ for the same value of n. This shift reflects the upward movement of $V_{\rm eff}(r)$. Accordingly, the probability of finding the electron near the nucleus, proportional to $|R_{n,l}|^2$, tends to 0.

With this shift away from the nucleus, the stabilization of the electron by the attraction of the nucleus diminishes. In contrast, in the case of $l = 0$ shown in Fig. 3.8(a), the instability due to this centrifugal effect is absent. Instead, the number of nodal planes at which $R(r)$ becomes 0 (cf. the small circles in Figs. 3.8(a) and (b)) is the largest for $l = 0$, when the three states of $l = 0, 1, 2$ for the same value of $n = 3$ are compared, which is another cause of instability.

The two different causes of instability happen to cancel out, resulting in the accidental degeneracy in the case of Coulombic potentials.

As is seen from Fig. 3.8, the number of nodal planes of the n-th energy state is $n - 1 - l$. This is derived more generally from what is called the oscillation theorem* which demands that the n-th eigenfunction counting from the $n = 1$ ground state must become null $n-1$ times if the eigenenergies are discrete. In the present case of a spherical system described by the total eigenfunction of $\phi(\boldsymbol{r}) = R(r)Y(\theta, \varphi)$, the angular component $Y(\theta, \varphi)$ becomes

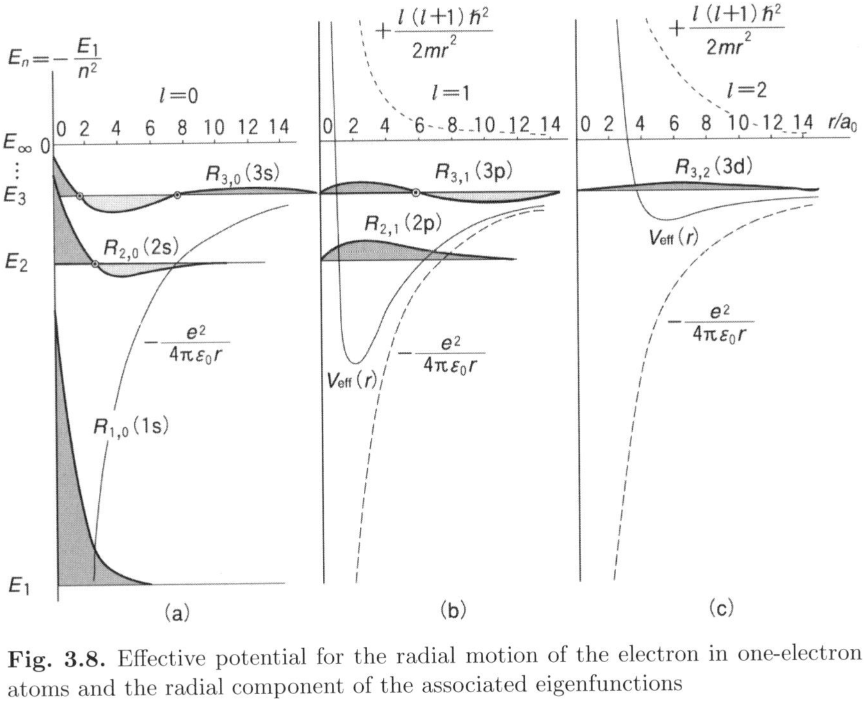

Fig. 3.8. Effective potential for the radial motion of the electron in one-electron atoms and the radial component of the associated eigenfunctions

null l times, as has been shown in Sect. 3.3.2. Therefore, the number of nodal planes in the $R(r)$ part for the n-th state is necessarily $n - 1 - l$. Conversely, in the n-th state, the quantum number l for is limited to $0, 1, 2, \ldots, (n - 1)$, as already mentioned in Sect. 3.2.3.

Thus far, the energy and the orbital angular momentum of a single electron in a Coulombic potential are accounted for. One thing left behind is the electron spin, which is a kind of angular momentum. The orbital angular momentum so far discussed is just one of the generalized angular momenta. In order to shed light on the new kind of angular momentum called the electron spin, the following section will be devoted to a discussion of the concept of generalized angular momentum in one-electron atoms.

3.4 Angular Momentum in One-Electron Atoms

(This section may be a little above the level of the book compared with the other sections. It may be put off a little while).

Up to this point it has been stated that there are two physical quantities having the dimension of angular momentum, i.e., the orbital angular momentum of the electron and the electron spin, and that the former has a classical analogue while the latter does not (cf. Sects. 1.3 and 1.4). These two quantities are relatively elusive in experimental detection compared with the energy of the electron which is spectroscopically probed with ease. However, the orbital angular momentum plays a fundamental role in, e.g., the classification of electronic states of spherically symmetric atoms, as well as in the group theoretical analysis of various problems of atoms and molecules. Likewise, the electron spin dictates the electronic state of many electron systems via the Pauli exclusion principle. Furthermore, the total angular momentum obtained by the addition of the orbital angular momentum and the electron spin is a gauge as decisive as the energy in systems where the total angular momentum is definable. It will be shown in this section that in quantum mechanics both the orbital angular momentum and the electron spin are defined by a common formula and are classified as the generalized angular momentum. However, the definition is necessarily abstract as it is unique to quantum mechanics with no analogue in classical mechanics. In this section I will divide the explanation into two parts. In the first half, the commutative relation (cf. Sect. 2.1.2) will be examined for the orbital angular momentum, which should be familiar by now, and then, in the latter half the generalized angular momentum will be presented.

3.4.1 The Definition of Orbital Angular Momentum by the Commutative Relationship

In the construction of the old quantum theory Bohr adopted Planck's hypothesis of quantization. To be specific, Bohr assumed a priori the condition in (1.10), which demands that the integral of the orbital angular momentum of an electron orbiting around a nucleus over one cycle of circulation must be limited to discrete values in units of the Planck constant. In contrast to this definition in the old quantum mechanics, the quantization according to Heisenberg's approach is based upon the uncertainty principle and starts with the following commutation relationship:

$$[\hat{r}_i, \hat{p}_j] \equiv (\hat{r}_i \hat{p}_j - \hat{p}_j \hat{r}_i) = i\hbar\delta_{ij}, \quad \delta_{ij} = 1\ (i = j), \quad \delta_{ij} = 0\ (i \neq j) \quad (3.24)$$

In the above relationship the three components of the position vector $\boldsymbol{r}$ and the momentum vector $\boldsymbol{p}$ of an electron are denoted as $\hat{r}_i, \hat{p}_j$ $(i, j = x, y, z)$, which represent the operations of $\hat{r}_i \to r_i\times$ and $\hat{p}_j \to \mathrm{i}\hbar\partial/\partial r_j$ as stated in Sect. 2.1.3. The notation $[\hat{r}_i, \hat{p}_j]$ stands for the commutator introduced in Sect. 2.1.12.

It is seen from (3.24) that $\hat{r}_x = x$, e.g., is commutative with $\hat{p}_y, \hat{p}_z$ but not with $\hat{p}_x$. This means that the position x and the momentum component along the y- or z-axis of an electron can be observed simultaneously with definite values but that the momentum component along the x-axis cannot be known simultaneously with the position x. Thus, (3.24) is another expression of the Heisenberg uncertainty principle mentioned in Sect. 1.2. It is obvious that the following relations in (3.25) for $\hat{\boldsymbol{r}}$ and $\hat{\boldsymbol{p}}$ hold identically irrespective of i and j.

$$[\hat{r}_i, \hat{r}_j] = 0, \quad [\hat{p}_i, \hat{p}_j] = 0 \quad (3.25)$$

Combining the above relations with those in (3.24) and referring to the vectorial operator for the orbital angular momentum $\hat{\boldsymbol{l}}$, which is written as $\hat{\boldsymbol{l}} = \hat{l}_x\boldsymbol{i} + \hat{l}_y\boldsymbol{j} + \hat{l}_z\boldsymbol{k}$ with $\hat{l}_x = y\hat{p}_z - z\hat{p}_y$, $\hat{l}_y = z\hat{p}_x - x\hat{p}_z$, and $\hat{l}_z = x\hat{p}_y - y\hat{p}_x$, the relationships shown in (3.26) are easily derived:

$$[\hat{l}_x, \hat{l}_y] = \mathrm{i}\hbar\hat{l}_z, \quad [\hat{l}_y, \hat{l}_z] = \mathrm{i}\hbar\hat{l}_x, \quad [\hat{l}_z, \hat{l}_x] = \mathrm{i}\hbar\hat{l}_y \quad (3.26)$$

These relationships can be concisely expressed also as $\hat{\boldsymbol{l}} \otimes \hat{\boldsymbol{l}} = \mathrm{i}\hbar\hat{\boldsymbol{l}}$ where the symbol $\otimes$ represents the operation of cross product of two vectors mentioned in Sect. 1.3. Furthermore, the following relations can also be immediately proven.

$$[\hat{\boldsymbol{l}}^2, \hat{l}_i] = 0\ (i = x, y, z), \quad \hat{\boldsymbol{l}}^2 \equiv \hat{l}_x^2 + \hat{l}_y^2 + \hat{l}_z^2 \quad (3.27)$$

When i is chosen as z, (3.27) means that there is a simultaneous eigenfunction f which satisfies the relations of $\hat{\boldsymbol{l}}^2 \times f = \mathrm{const} \times f$ and $\hat{l}_z \times f = \mathrm{const} \times f$ as was explained in Sect. 2.1.12. From (3.11), (3.14), and (3.22) we are already aware that this function f is equal to one of the spherical surface harmonics $Y_{l,m}$. Equation (3.27) predicts the existence of a simultaneous eigenfunction common to $\hat{\boldsymbol{l}}^2$ and $\hat{l}_z$, irrespective of the explicit form of the function.

So far it has been confirmed that the two operators $\hat{\boldsymbol{l}}^2$ and $\hat{l}_i$ concerned with the orbital angular momentum are commutative with common eigenfunctions. Next, the definition of the generalized angular momentum will be explored.

Problem

3.2 Confirm the relations in (3.26) and (3.27). As noticed in Sect. 1.2, in confirming a relation between operators, always put an arbitrary function to the right of the operators and compare the result of the operations on both sides of the relation.

3.3 As is seen from (3.24), the dimensions of both sides of the identity are equal to the dimension of the angular momentum. In such a case the coordinate $\boldsymbol{r}$ and the momentum $\boldsymbol{p}$ may be said to be conjugate with regard to the angular momentum. Like the pair of $\boldsymbol{r}$ and $\boldsymbol{p}$, the energy E, whose operator representation is equal to $\mathrm{i}\hbar\partial/\partial t$ as given in (3.17), and the time t are also conjugate to each other in the above sense. Show that the commutation relation $[\mathrm{i}\hbar\partial/\partial t, t] = \mathrm{i}\hbar$ holds good for the pair of E and t also. This relationship also is often used to express the uncertainty principle.

3.4.2 The Generalized Angular Momentum

Suppose that the generalized angular momentum mentioned in the beginning of Sect. 3.4 can be defined in such a way as to comprise the familiar orbital angular momentum $\hat{\boldsymbol{l}}$. Hereafter, the generalized angular momentum will be designated by $\hat{\boldsymbol{j}}$ to differentiate it from $\hat{\boldsymbol{l}}$. Assume that $\hat{\boldsymbol{j}}$ satisfies the same relations as in (3.26) for $\hat{\boldsymbol{l}}$.

$$[\hat{j}_x, \hat{j}_y] = \mathrm{i}\hbar\hat{j}_z, \quad [\hat{j}_y, \hat{j}_z] = \mathrm{i}\hbar\hat{j}_x, \quad [\hat{j}_z, \hat{j}_x] = \mathrm{i}\hbar\hat{j}_y \tag{3.28}$$

These relations may be regarded as the defining equations of the generalized angular momentum. The three equations may also be represented compactly as $\hat{\boldsymbol{j}} \otimes \hat{\boldsymbol{j}} = \mathrm{i}\hbar\hat{\boldsymbol{j}}$.

Defining the square of $\hat{\boldsymbol{j}}$ by $\hat{\boldsymbol{j}}^2 = \hat{j}_x^2 + \hat{j}_y^2 + \hat{j}_z^2$, the relation $[\hat{\boldsymbol{j}}^2, \hat{j}_i] = 0$ with $i = x, y, z$ is obtained from (3.28). This relation is exactly the same as for the orbital angular momentum in (3.27).

At this stage the artificial operators defined by (3.29) below will be introduced. They are artificial in the sense that they do not correspond to any physical quantity but are utilized for the sake of computational convenience. Thus, there is no need to worry about the fact that they are not Hermitian (cf. the discussion in Sect. 2.1.8).

$$\hat{j}_{+1} = -\frac{1}{\sqrt{2}}(\hat{j}_x + \mathrm{i}\hat{j}_y), \quad \hat{j}_0 = \hat{j}_z, \quad \hat{j}_{-1} = \frac{1}{\sqrt{2}}(\hat{j}_x - \mathrm{i}\hat{j}_y) \tag{3.29}$$

The newly introduced operators $\hat{j}_{+1}$ and $\hat{j}_{-1}$ are called step operators or ladder operators. The coefficients $\mp 1/\sqrt{2}$ originate from the fact that the normalization constant of the eigenfunction for the orbital angular momentum in Sect. 3.2.3 has been chosen to be consistent with the normalization constant of the functions in (3.15). Alternatively, the definition $\hat{j}_{\pm 1} \equiv \hat{j}_x \pm \mathrm{i}\hat{j}_y$ may also be found in other textbooks.

With the use of $\hat{j}_{+1}$, $\hat{j}_0$, and $\hat{j}_{-1}$, the following relationships are obtained for $\hat{j}_x$, $\hat{j}_y$ and $\hat{j}_z$.

$$\hat{j}_x = -\frac{1}{\sqrt{2}}(\hat{j}_{+1} - \hat{j}_{-1}), \quad \hat{j}_y = \frac{\mathrm{i}}{\sqrt{2}}(\hat{j}_{+1} + \hat{j}_{-1}), \quad \hat{j}_z = \hat{j}_0 \tag{3.30}$$

According to quantum mechanics, it can be derived solely from the commutation relation of $\hat{\boldsymbol{j}}^2, \hat{j}_x, \hat{j}_y, \hat{j}_z, \hat{j}_{+1}$, and $\hat{j}_{-1}$ that the operators $\hat{\boldsymbol{j}}^2$ and $\hat{j}_z$ have a common eigenfunction $f_{j,m}$ as in (3.31) and (3.32) below and that the step operators obey the relation in (3.33). The derivation of the odd looking relationship of (3.33) is straightforward but is a little lengthy so it is omitted here and left to textbooks of elementary quantum mechanics.

$$\hat{\boldsymbol{j}}^2 f_{j,m} = j(j+1)\hbar^2 f_{j,m} \tag{3.31}$$

$$\hat{j}_z f_{j,m} = m\hbar f_{j,m} \tag{3.32}$$

$$\hat{j}_{\pm 1} f_{j,m} = \mp\sqrt{1/2}\sqrt{j(j+1) - m(m\pm 1)}\hbar f_{j,m\pm 1} \tag{3.33}$$

Furthermore, by substituting (3.33) into (3.30), the following results are obtained which show the effect of the operation of $\hat{j}_x$ and $\hat{j}_y$ upon the eigenfunctions.

$$\begin{aligned}\hat{j}_x f_{j,m} &= \frac{1}{2}\sqrt{j(j+1) - m(m+1)}\hbar f_{j,m+1} \\ &+\frac{1}{2}\sqrt{j(j+1) - m(m-1)}\hbar f_{j,m-1}\end{aligned} \tag{3.34}$$

$$\begin{aligned}\hat{j}_y f_{j,m} &= \frac{-\mathrm{i}}{2}\sqrt{j(j+1) - m(m+1)}\hbar f_{j,m+1} \\ &+\frac{\mathrm{i}}{2}\sqrt{j(j+1) - m(m-1)}\hbar f_{j,m-1}\end{aligned} \tag{3.35}$$

From (3.33) the relations $\hat{j}_{+1} f_{j,j} = 0$ and $\hat{j}_{-1} f_{j,-j} = 0$ are required from which the relation $j \geq |m|$ is derived automatically. This relationship is analogous to the relationship $l \geq |m|$ for the orbital angular momentum as mentioned in Sect. 3.2.3. As a result, the number of possible values of m in (3.32) is restricted to $2j+1$, which corresponds to $2l+1$ in the case of the orbital angular momentum (cf. Sect. 3.2.3).

What is most important is that in the case of the orbital angular momentum the quantum number l is limited to $l = 0, 1, 2, \ldots, (n-1)$ as shown in Sect. 3.2.3. Therefore, the value of $2l+1$ must be an odd integer. In contrast, in the case of the generalized angular momentum there is no such limitation on $2j+1$, i.e., both odd and even integers are allowed. When even, j must be a fractional half integer $1/2, 3/2, 5/2, \ldots$ and the possible values for m run from the maximum $m = j$ (> 0) to the minimum $m = -j$ as

$m = j, j-1, \ldots, -(j-1), -j$. For example, when $j = 3/2$, the possible values for m are $m = \pm 3/2$ and $m = \pm 1/2$. The newly allowed fractional half integers for j are the quantum numbers for the spins of the electron and nuclides consisting of nucleons (i.e., protons and neutrons) of an odd number. As for the electron spin, nature has chosen a value of $j = 1/2$ by chance, so that there are only two possible states corresponding to $m = \pm 1/2$. For the electron spin, it is customary to use the symbols s and m_s in place of j and m. Similarly, for the nuclear spin, the symbols I and m_I are preferred to j and m, although in this book we will not deal with nuclear spin.

Before closing this section I will repeat that the quantum number l for the orbital angular momentum is limited to $n - 1$ by the boundary conditions proper to the spherically symmetric Coulombic potential of one-electron atoms. The corresponding quantum number s for the electron spin, on the other hand, is limited to just $s = 1/2$ by nature. As for the general spherical surface harmonics $Y_{l,m}$ satisfying an eigenvalue equation of the type of (3.11), the index l runs as $l = 0, 1, 2, \ldots, \infty$ and the whole set of $Y_{l,m}$ guarantees the completeness of the eigenfunctions.

3.4.3 Eigenfunctions of the Electron Spin

The eigenfunctions of the orbital angular momentum of electron are given by the spherical surface harmonics $Y_{l,m}(\theta, \varphi)$ as a function of the angular coordinates θ and φ. These functions correspond to the functions $f_{j,m}$ appearing in the preceding section in the discussion of the generalized angular momentum. In this section the eigenfunctions of the electron spin, which belongs to the generalized angular momentum, will be examined. As a generalized angular momentum, the electron spin operators must obey the set of defining relations in (3.28), and the eigenfunctions for the electron spin must satisfy the eigenvalue equations (3.31) and (3.32) resulting from the definition. Hereafter, the symbols for the generalized angular momentum in the preceding section will be replaced with $\hat{\boldsymbol{j}} \to \hat{\mathbf{s}}$, $j \to s$, $m \to m_s$, $f_{j,m} \to \varphi_{s,m_s}$. This is just to remind us that we are dealing with the electron spin exclusively. Thus, the eigenvalue equations corresponding to (3.31) and (3.32) are written as follows.

$$\hat{\mathbf{s}}^2 \varphi_{s,m_s} = s(s+1)\hbar^2 \varphi_{s,m_s} = \frac{3}{4}\hbar^2 \varphi_{s,m_s} \tag{3.36}$$

$$\hat{s}_z \varphi_{s,m_s} = m_s \hbar \varphi_{s,m_s} = \pm\frac{1}{2}\hbar \varphi_{s,m_s} \tag{3.37}$$

The two states corresponding to the eigenvalues of $m_s\hbar$ equal to $+\hbar/2$ and $-\hbar/2$ will be called the spin-up state and spin-down state, respectively. To describe the state of an electron it is necessary to specify which spin state is assumed by the electron. Thus, in addition to the three positional coordinates, a fourth degree of freedom for the differentiation of the spin state is needed.

This degree of freedom will be called the spin coordinate and will be denoted by σ. The coordinate is allowed to assume a value of either $\sigma = +1/2$ or $\sigma = -1/2$ depending upon whether the electron is in the spin-up or down state, respectively. In other words, the values of $m_s = +1/2$ and $m_s = -1/2$, which are obtained when the eigenvalue $m_s\hbar$ in (3.37) is measured in units of $\hbar$, are identified as the assumed values for the spin coordinate. It may appear somewhat odd to use the *eigenvalue* m_s of a physical quantity, i.e., the electron spin as the value for the spin *coordinate* σ to be assumed by an electron. However, it is to be accepted as is because the electron spin is a purely quantum mechanical quantity beyond physical understanding.

Once this is accepted, the spin eigenfunctions $\varphi_{s,m_s=\pm 1/2}$ and $\varphi_{s,m_s=-1/2}$ are denoted concisely as α and β, respectively, and the eigenvalue equations in (3.37) can be written as $\hat{s}_z\alpha = +(\hbar/2)\alpha$ and $\hat{s}_z\beta = -(\hbar/2)\beta$. That is to say, the eigenfunction α is 1 when the spin coordinate σ assumes the value of $+1/2$ and is 0 when σ equal to $-1/2$. Likewise, the eigenfunction β is 1 when $\sigma = -1/2$ and is 0 when $\sigma = +1/2$. In short, it may be expressed as $\alpha(+1/2) = 1$, $\alpha(-1/2) = 0$ and $\beta(+1/2) = 0$, $\beta(-1/2) = 1$, or vectorially, as in (3.38), where the equal sign (=) simply means that α and β may sometimes be designated as $\binom{1}{0}$ and $\binom{0}{1}$, respectively.

$$\varphi_{s,m_s=+1/2} \equiv \alpha = \begin{pmatrix} 1 \\ 0 \end{pmatrix}, \quad \varphi_{s,m_s=-1/2} \equiv \beta = \begin{pmatrix} 0 \\ 1 \end{pmatrix} \tag{3.38}$$

A brief comment was made in Sect. 1.4.2 regarding the combining of the solution of the Schrödinger equation (concerned only with the spatial coordinates and the time) with the electron spin. The comment is repeated here by saying that a simple product of the solution of $\psi(\boldsymbol{r}, t)$ in Sect. 1.4.2 with the spin factor of α or β can be regarded as the total solution in the Schrödinger-Pauli treatment. However, since we will not be concerned with time-dependent phenomena such as the transition between different energy states induced by electromagnetic waves, we can reduce $\psi(\boldsymbol{r}, t)$ to $\phi(\boldsymbol{r})$ as argued in Sect. 2.1.2 and consider the combination of $\phi(\boldsymbol{r})$ with the spin functions α and β. The result of this combination is denoted as $\phi(\boldsymbol{r}\sigma)$ and will be called the spin orbital hereafter. From the above statements the spin orbitals with spin states α and β can be written, respectively, as $\phi(\boldsymbol{r}\sigma) = \phi(\boldsymbol{r})\alpha(\sigma)$ and $\phi(\boldsymbol{r}\sigma) = \phi(\boldsymbol{r})\beta(\sigma)$.

Since the spin state is described only by α and β which are the eigenfunctions of the eigenvalue equations of (3.31) and (3.32), any spin orbital $\phi(\boldsymbol{r}\sigma)$ can be expressed by a linear combination of the two eigenfunctions $\alpha(\sigma)$ and $\beta(\sigma)$ by virtue of the completeness of eigenfunctions (cf. Sect. 2.1.10). The coefficients of the linear combinations should not depend on the spin coordinate σ, but they should be functions of the spatial coordinate of the electron. Thus, when these coefficients are written as $\phi_\alpha(\boldsymbol{r})$ and $\phi_\beta(\boldsymbol{r})$, the general expression for any spin orbital becomes $\phi(\boldsymbol{r}\sigma) = \phi_\alpha(\boldsymbol{r})\alpha(\sigma) + \phi_\beta(\boldsymbol{r})\beta(\sigma)$. The physical meaning of the coefficients $\phi_\alpha(\boldsymbol{r})$ and $\phi_\beta(\boldsymbol{r})$ is such that the squares of their absolute values multiplied by the spatial volume element $dv = dxdydz$ represent the probability of finding an electron either in the spin-up or in the spin-down state in that volume element. Since the spin state of the electron must be either one of the two, the integral over the whole *spatial space* of the probability must be unity, i.e., $\int \left[|\phi_\alpha(\boldsymbol{r})|^2 + |\phi_\beta(\boldsymbol{r})|^2\right] dv = 1$. It is apparent that the general spin orbital, $\phi(\boldsymbol{r}\sigma) = \phi_\alpha(\boldsymbol{r})\alpha(\sigma) + \phi_\beta(\boldsymbol{r})\beta(\sigma)$, is not an eigenfunction of the operator $\hat{s}_z$. Therefore, when an electron is in the state described by the above $\phi(\boldsymbol{r}\sigma)$, the spin state is not uniquely determined. However, it seldom occurs that such a general expression is necessary in the study of chemical bonding, and in most cases the spin orbital is simply described as $\phi(\boldsymbol{r}\sigma) = \phi_\alpha(\boldsymbol{r})\alpha(\sigma)$ or $\phi(\boldsymbol{r}\sigma) = \phi_\beta(\boldsymbol{r})\beta(\sigma)$.

As for the integral of $|\alpha(\sigma)|^2$ and $|\beta(\sigma)|^2$ over the whole *space of the spin coordinate*, it is replaced by the sums $\int |\alpha(\sigma)|^2 d\sigma = |\alpha(+1/2)|^2 + |\alpha(-1/2)|^2 = 1$ and $\int |\beta(\sigma)|^2 d\sigma = |\beta(+1/2)|^2 + |\beta(-1/2)|^2 = 1$ because the eigenfunction in the spin space assumes only the four values of $\alpha(+1/2) = 1$, $\alpha(-1/2)$, $\beta(+1/2) = 0$, and $\beta(-1/2) = 1$. The integrals above show that the eigenfunctions $\alpha(\sigma)$ and $\beta(\sigma)$ are normalized. Furthermore, they are orthogonal because an integral such as $\int \alpha^*(\sigma)\beta(\sigma)d\sigma$, when replaced with $\alpha^*(+1/2)\beta(+1/2) + \alpha^*(-1/2)\beta(-1/2)$, is equal to 0. This orthogonality is a natural consequence of the eigenvalue equations (3.36) and (3.37) because $\alpha(\sigma)$ and $\beta(\sigma)$ correspond to the eigenfunction φ_{s,m_s} in the two equations. In this way it is shown that the spin orbital $\phi(\boldsymbol{r}\sigma)$ is normalized with regard to both the positional and the spin coordinates to obtain $\int\int |\phi(\boldsymbol{r}\sigma)^2 dv d\sigma = 1$.

The spin-up and spin-down states of the electron are frequently compared to the left and right self-revolutions of an electron. However, since the electron spin is to be described in the spin space only and not in the positional space, this should be taken as a mere analogy.

3.4.4 Electron Spin Operators

(This section is for reference and can be skipped. However, some of the relationships below will be used in Sect. 7.2.2).

As shown in the preceding Sects. 3.4.2 and 3.4.3, the electron spin is a generalized angular momentum. As in Sect. 3.4.3, the symbols in (3.31), (3.32), (3.34), and (3.35) are replaced by $j \to s$, $m \to m_s$, and $f_{j,m} \to \varphi_{s,m_s}$. The quantum numbers of the electron spin are restricted to $s = 1/2$ and $m_s = \pm 1/2$, and the result of the respective operations of $\hat{s}_{+1}$ and $\hat{s}_{-1}$ upon the eigenfunctions of $\varphi(s, m_s = +1/2) \equiv \alpha$ and $\varphi(s, m_s = -1/2) \equiv \beta$ must be 0 because the operations correspond to the general requirements of $\hat{j}_{+1} f_{j,j} = 0$ and $\hat{j}_{-1} f_{j,-j} = 0$, as stated in connection with (3.33). As a result, the following relationships are derived.

$$\hat{s}_x\alpha = \frac{1}{2}\sqrt{s(s+1) - m_s(m_s-1)}\hbar\beta = \frac{1}{2}\hbar\beta \qquad (3.39.\text{i})$$

$$\hat{s}_x\beta = \frac{1}{2}\sqrt{s(s+1) - m_s(m_s+1)}\hbar\alpha = \frac{1}{2}\hbar\alpha \quad (3.39.\text{ii})$$

$$\hat{s}_y\alpha = +\frac{\mathrm{i}}{2}\sqrt{s(s+1) - m_s(m_s-1)}\hbar\beta = +\frac{\mathrm{i}}{2}\hbar\beta \quad (3.39.\text{iii})$$

$$\hat{s}_y\beta = -\frac{\mathrm{i}}{2}\sqrt{s(s+1) - m_s(m_s+1)}\hbar\beta = -\frac{\mathrm{i}}{2}\hbar\alpha \quad (3.39.\text{iv})$$

$$\hat{s}_z\alpha = m_s\alpha = \frac{1}{2}\hbar\alpha \quad (3.39.\text{v})$$

$$\hat{s}_z\beta = m_s\hbar\beta = -\frac{1}{2}\hbar\beta \quad (3.39.\text{vi})$$

$$\hat{\boldsymbol{s}}^2\alpha = s(s+1)\hbar^2\alpha = \frac{3}{4}\hbar^2\alpha \quad (3.39.\text{vii})$$

$$\hat{\boldsymbol{s}}^2\beta = s(s+1)\hbar^2\beta = \frac{3}{4}\hbar^2\beta \quad (3.39.\text{viii})$$

The pair of (v) and (vii) and that of (vi) and (viii) demonstrate that the eigenfunctions α and β are simultaneous eigenfunctions of the two operators $\hat{s}_z$ and $\hat{\boldsymbol{s}}^2$ and that the eigenvalues for α are $s = 1/2$ and $m_s = 1/2$ and those for β are $s = 1/2$ and $m_s = -1/2$. On the other hand, α and β are not eigenfunctions of the operators $\hat{s}_x$ and $\hat{s}_y$. This means that in the states α and β the eigenvalues of $\hat{s}_z$ and $\hat{\boldsymbol{s}}^2$ are definitely known while the values for $\hat{s}_x$ and $\hat{s}_y$ are undetermined.

Although we will not go too far into it here, it was mentioned that the relationships in (3.39.i) through (3.39.viii) for the operators are replaced in matrix mechanics by the relationships of matrices and vectors. In order to deal with the electron spin via matrix mechanics Pauli introduced 2×2 matrices called the Pauli matrix* and the 8 relations above are replaced by the relations below. This approach was explored by Dirac, who founded the relativistic quantum mechanics*.

$$\begin{pmatrix} 0 & 1/2 \\ 1/2 & 0 \end{pmatrix}\begin{pmatrix} 1 \\ 0 \end{pmatrix} = \frac{1}{2}\hbar\begin{pmatrix} 0 \\ 1 \end{pmatrix} \quad (3.40.\text{i})$$

$$\begin{pmatrix} 0 & 1/2 \\ 1/2 & 0 \end{pmatrix}\begin{pmatrix} 0 \\ 1 \end{pmatrix} = \frac{1}{2}\hbar\begin{pmatrix} 1 \\ 0 \end{pmatrix} \quad (3.40.\text{ii})$$

$$\begin{pmatrix} 0 & -\mathrm{i}/2 \\ \mathrm{i}/2 & 0 \end{pmatrix}\begin{pmatrix} 1 \\ 0 \end{pmatrix} = +\frac{\mathrm{i}}{2}\hbar\begin{pmatrix} 0 \\ 1 \end{pmatrix} \quad (3.40.\text{iii})$$

$$\begin{pmatrix} 0 & -\mathrm{i}/2 \\ \mathrm{i}/2 & 0 \end{pmatrix}\begin{pmatrix} 0 \\ 1 \end{pmatrix} = -\frac{\mathrm{i}}{2}\hbar\begin{pmatrix} 1 \\ 0 \end{pmatrix} \quad (3.40.\text{iv})$$

$$\begin{pmatrix} 1/2 & 0 \\ 0 & -1/2 \end{pmatrix}\begin{pmatrix} 1 \\ 0 \end{pmatrix} = \frac{1}{2}\hbar\begin{pmatrix} 1 \\ 0 \end{pmatrix} \quad (3.40.\text{v})$$

$$\begin{pmatrix} 1/2 & 0 \\ 0 & -1/2 \end{pmatrix} \begin{pmatrix} 0 \\ 1 \end{pmatrix} = -\frac{1}{2}\hbar \begin{pmatrix} 0 \\ 1 \end{pmatrix} \tag{3.40.vi}$$

$$\begin{pmatrix} 3/4 & 0 \\ 0 & 3/4 \end{pmatrix} \begin{pmatrix} 1 \\ 0 \end{pmatrix} = \frac{3}{4}\hbar^2 \begin{pmatrix} 1 \\ 0 \end{pmatrix} = \frac{1}{2}\left(\frac{1}{2}+1\right)\hbar^2 \begin{pmatrix} 1 \\ 0 \end{pmatrix} \tag{3.40.vii}$$

$$\begin{pmatrix} 3/4 & 0 \\ 0 & 3/4 \end{pmatrix} \begin{pmatrix} 0 \\ 1 \end{pmatrix} = \frac{3}{4}\hbar^2 \begin{pmatrix} 0 \\ 1 \end{pmatrix} = \frac{1}{2}\left(\frac{1}{2}+1\right)\hbar^2 \begin{pmatrix} 0 \\ 1 \end{pmatrix} \tag{3.40.viii}$$

The last four equations demonstrate that the two spin functions $\alpha = \binom{1}{0}$ and $\beta = \binom{0}{1}$ are simultaneous eigenfunctions of the two operators $\hat{s}_z$ and $\hat{\boldsymbol{s}}^2$ with the respective eigenvalues of $m_s = +1/2$, $s = 1/2$ and $m_s = -1/2$, $s = 1/2$.

The operators $\hat{s}_x$, $\hat{s}_y$, $\hat{s}_z$, and $\hat{\boldsymbol{s}}^2$ should satisfy the commutation relations in (3.28) because they belong to the generalized angular momentum. The commutation relations can be confirmed by replacing the operators with the corresponding matrices and executing multiplications among the matrices.

4 Multi-Electron Atoms: The Building Blocks that Produce the Tremendous Variety of Molecules

Atoms comprising more than one electron are much more complicated to analyze than the simple one-electron atom. The major reasons for the difficulties are as follows.

(1) The electrostatic potential felt by an electron is the sum of attractive potentials due to the central nucleus and the repulsive potentials due to all the other electrons. As a result, the potential of any single electron concerned cannot be expressed in such a simple form as $V(r) = -Ze^2/(4\pi\varepsilon_0 r)$ any more; as a result, the eigenvalue and the eigenfunction cannot be solved for analytically.

(2) Electrons in the multi-electron atom are indistinguishable from each other, and any two electrons cannot be in one and the same state on account of the Pauli exclusion principle. Some means must be found to meet these a priori demands.

(3) The orbital angular momentum and the electron spin of a single electron are vectorial operators. Accordingly, in multi-electron atoms the vectors of each electron must be summed vectorially. Moreover, the total angular momentum of the electrons thus summed up must be quantized properly to obtain permissible energy states of the multi-electron atom.

The basic assumption in the treatment of multi-electron atoms is that the energy eigenfunctions of the atoms can be described by assigning a single spin orbital $\phi(\boldsymbol{r}\sigma) = \phi(\boldsymbol{r})\alpha(\sigma)$ or $\phi(\boldsymbol{r}\sigma) = \phi(\boldsymbol{r})\beta(\sigma)$ to each electron. Then, monitoring the total energy of the atoms, the best approximate eigenfunctions and eigenvalues are sought through the application of techniques such as the calculus of variation and the perturbation method to find a minimum value of the total energy.

In this book I will restrict the discussion of the subject of multi-electron atoms to those topics relevant to the understanding of chemical bonding in line with the statement in the preface.

4.1 A Requirement Imposed on the Energy Eigenfunction of Multi–Electron Atoms: The Pauli Exclusion Principle

By a twist of nature, the electron happens to be a fermion. Any fermion must obey the Pauli exclusion principle.

4.2 The Energy of Multi-Electron Atoms: A Qualitative Consideration and the Concept of Electronic Shell Structures

The energy of an electron in a non-Coulombic spherically symmetric potential depends not only on the principal quantum number but also on the azimuthal quantum number, different from the case of the electron in a one-electron atom discussed in Chap. 3. Thus, the electronic energy state is differentiated by these two quantum numbers.

4.3 The Angular Momentum of Multi-Electron Atoms: The Total Angular Momentum as a Physical Quantity to Characterize the Electronic State of Multi-Electron Atoms
The total angular momentum is as important as the energy for the description of multi-electron atoms. This situation is the same as that in one-electron atoms.

4.4 An Example of Solving the Problem of Multi-Electron Atoms by the Hartree-Fock Method: For the Case of Ne
The first step for the quantitative solution of the electronic state in multi-electron atoms is to apply the so-called SCF method. The procedure is essentially the same as that for polyatomic molecules in Chap. 8. To avoid redundancy, only the results of the application of this method to the Ne atom are presented in this section. The reader is solicited to return to this chapter after grasping the principles of the SCF method worked out in Chap. 8.

4.1 A Requirement Imposed on the Energy Eigenfunction of Multi-Electron Atoms: The Pauli Exclusion Principle

Let me first explain the Pauli exclusion principle as a very general principle governing any multi-electron system, not only multi-electron atoms to be dealt with in this chapter, but also polyatomic molecules in Chaps. 7 and 8. The eigenvalue equation for the energy of multi-electron systems is expressed by the following extension of (2.6), used for single-electron systems:

$$\begin{aligned}
&\hat{H}\Phi(\boldsymbol{r}_1\sigma_1, \boldsymbol{r}_2\sigma_2, \ldots, \boldsymbol{r}_N\sigma_N) \\
&= \left[\sum_{i=1}^{N} -\frac{\hbar^2}{2m}\Delta_i + V\right]\Phi(\boldsymbol{r}_1\sigma_1, \boldsymbol{r}_2\sigma_2, \ldots, \boldsymbol{r}_N\sigma_N) \\
&= E\Phi(\boldsymbol{r}_1\sigma_1, \boldsymbol{r}_2\sigma_2, \ldots, \boldsymbol{r}_N\sigma_N) \qquad (4.1)
\end{aligned}$$

In the above the symbol for the eigenfunction has been changed from the small letter ϕ in (2.6) for a single-electron system to the capital letter Φ to emphasize that the system under consideration is a function of many variables.

The first term of the Hamiltonian in the square brackets of (4.1) represents the total kinetic energy of N electrons whose mass m is slightly reduced compared with the mass of the electron at rest m_e (cf. the note at the end of Sect. 3.1). The second term in the brackets represents the total potential energy operator, which is a function of the coordinates of all the electrons. Leaving this potential term as is for a while, we look into the eigenfunction first.

At the moment, nothing is known about the dependence of Φ upon the positional coordinate $\boldsymbol{r}$ and the spin coordinate σ of each electron. As a trial, however, let us assume that each electron behaves independently of all the other electrons and that, to each electron, a spin orbital $\phi(\boldsymbol{r}\sigma)$ discussed in Sect. 3.4.3 is assignable. This is the one-electron approximation, by which the function for the total system is expressed as a simple product of each spin orbital. This approximation is analogous to the theory of probability which states that the total probability is given by the product of the probability of individual events which are independent of each other. Thus, the total eigenfunction Φ is written as $\Phi(\boldsymbol{r}_1\sigma_1, \boldsymbol{r}_2\sigma_2, \ldots, \boldsymbol{r}_N\sigma_N) = \phi_{k_1}(\boldsymbol{r}_1\sigma_1)\phi_{k_2}(\boldsymbol{r}_2\sigma_2)\ldots\phi_{k_N}(\boldsymbol{r}_N\sigma_N)$ where each factor ϕ on the right-hand side represents a normalized one-electron function called the spin orbital in Sect. 3.4.3. The subscript on the coordinate of a single electron $\boldsymbol{r}\sigma$ is to differentiate each electron and the subscript on the spin orbital is to distinguish the functions ϕ. The above equation will be denoted concisely as $\Phi(1, 2, \ldots, N) = \phi_{k_1}(1)\phi_{k_2}(2)\ldots\phi_{k_N}(N)$.

Now, suppose that the i-th and the j-th electrons are exchanged. Since all the electrons are equivalent, the Hamiltonian $\hat{H}$ in (4.1) does not change with this exchange, whatever the form of the potential term V may be. Therefore, if the function $\Phi(1, 2, \ldots, i, \ldots, j, \ldots, N)$ satisfies (4.1), then, the function $\Phi(1, 2, \ldots, j, \ldots, i, \ldots, N)$ also should satisfy the same equation. This conclusion is tantamount to saying that the above two functions differ only by a constant multiplication factor under the proviso that the two are not degenerate (cf. Sects. 2.1.4 and 2.1.7). Thus, we have the relationship, $\Phi(1, 2, \ldots, j, \ldots, i, \ldots, N) = \text{const} \times \Phi(1, 2, \ldots, i, \ldots, j, \ldots, N)$. If we regard the above exchange as an action of an operator $\hat{O}$ on $\Phi(1,2,\ldots,i,\ldots,j,\ldots,N)$, the above relationship can be written as $\hat{O}\Phi(1, 2, \ldots, i, \ldots, j, \ldots, N) = \Phi(1, 2, \ldots, i, \ldots, j, \ldots, N) = \text{const} \times \Phi(1, 2, \ldots, j, \ldots, i, \ldots, N)$, which implies that the constant is an eigenvalue of the operator $\hat{O}$. Repeated application of $\hat{O}$ to the above equation gives the relationship, $\hat{O}\hat{O}\Phi(1, 2, \ldots, i, \ldots, j, \ldots, N) = \text{const} \times \hat{O}\Phi(1, 2, \ldots, i, \ldots, j, \ldots, N) = (\text{const})^2 \times \Phi(1, 2, \ldots, i, \ldots, j, \ldots, N)$. Since the repetition of the action of $\hat{O}$ recovers the system to the original state, we have $(\text{const})^2 = 1$, hence $\text{const} = \pm 1$. This is equivalent to saying that the eigenfunction of the whole system $\Phi(1, 2, \ldots, N)$ remains unchanged or changes its sign with the exchange of $i \leftrightarrow j$.

Knowing this behavior of the total eigenfunction with the exchange of electrons, we attempt to find the solution which is guaranteed to remain unchanged or to change the sign with the exchange. Obviously, each of the two linear combinations, $\Phi(1, 2, \ldots, i, \ldots, j, \ldots, N) \pm \Phi(1, 2, \ldots, j, \ldots, i, \ldots, N)$, satisfies one of the above requirements. Both of them satisfy (4.1) because

$\hat{H}$ is a linear operator as shown in Sect. 2.1.4 so that any linear combination of constituent functions is acceptable from that point of view. In the case of the plus-combination the sum remains unchanged while the difference for the minus-combination changes its sign.

According to the conclusion reached inductively from any experiment tested so far, nature allows only two types of systems consisting of identical particles. Being identical here means that each particle is indistinguishable under any condition. In one of the two types above, the function describing the system always remains unchanged with the exchange of any two particles. In the other, the function changes sign for any such exchange. Thus, the multi-electron atom considered above is just one example of such a system of identical particles. A particle in the system which changes sign is called a Fermi particle or fermion, while a particle in the other system is called Bose particle or boson.

It happens that the electron is a Fermi particle. Any Fermi particle is associated with a spin quantum number of fractional half integers as stated in Sect. 3.4.2. Like the electron, the proton and the neutron are also Fermi particles. In contrast, a Bose particle is associated with integer spin quantum numbers. Discourse on these two types of particles will not be continued in this book, however.

For the problem of N-electron atoms, the number of ways of allotting each electron to N-spin orbitals ϕ is equal to $N!$. The indistinguishability of individual electrons requires that all possible permutations of electron distribution in the occupation of spin orbitals be considered equally. This is realized by a linear combination of the results of all the permutations with equal weights.

By the way, any permutation in a system of N elements can be attained by a repetition of exchanges of two elements. One can start by exchanging any two elements, but the number of binary exchanges necessary to attain a desired permutation is fixed to be either odd or even (including 0).

A concrete example of this statement is as follows. Consider a three-electron system. Assume that the initial distribution was $\phi_{k_1}(1)\phi_{k_2}(2)\phi_{k_3}(3)$ and that the final permutation is, e.g., $\phi_{k_1}(2)\phi_{k_2}(3)\phi_{k_3}(1)$. One way to obtain that result is through the exchange of electrons 1 and 2 followed by the exchange of 1 and 3, i.e., $\phi_{k_1}(1)\phi_{k_2}(2)\phi_{k_3}(3) \to \phi_{k_1}(2)\phi_{k_2}(1)\phi_{k_3}(3) \to \phi_{k_1}(2)\phi_{k_2}(3)\phi_{k_3}(1)$. Since the sign changes for each exchange, the sign for the final result obtained by two exchanges must be $(-1)^2$. The same sign is obtained if the first exchange was between 1 and 3, or between 2 and 3. The result of the above double exchange can also be obtained by the following quadruple exchange $\phi_{k_1}(1)\phi_{k_2}(2)\phi_{k_3}(3) \to \phi_{k_1}(3)\phi_{k_2}(2)\phi_{k_3}(1) \to \phi_{k_1}(3)\phi_{k_2}(1)\phi_{k_3}(2) \to \phi_{k_1}(1)\phi_{k_2}(3)\phi_{k_3}(2) \to \phi_{k_1}(2)\phi_{k_2}(3)\phi_{k_3}(1)$. A permutation is called 'even' or 'odd' depending on the number of exchanges. If we assign a plus sign to all the even permutations and a minus to the odd permutations, and sum them, then, the result becomes a linear combination of 3! products of the spin orbitals for the three electrons as follows, $\phi_{k_1}(1)\phi_{k_2}(2)\phi_{k_3}(3) + \phi_{k_1}(2)\phi_{k_2}(3)\phi_{k_3}(1) + \phi_{k_1}(3)\phi_{k_2}(1)\phi_{k_3}(2) - \phi_{k_1}(3)\phi_{k_2}(2)\phi_{k_3}(1) - \phi_{k_1}(1)\phi_{k_2}(3)\phi_{k_3}(2) - \phi_{k_1}(2)\phi_{k_2}(1)\phi_{k_3}(3)$.

The above sum describes the three Fermi particle system because the exchange of any pair of the three electrons changes the sign of the sum, which is required for an eigenfunction of the system. The above sum happens to be the same in value as the following determinant, called the Slater determinant.

$$\begin{vmatrix} \phi_{k_1}(1) & \phi_{k_2}(1) & \phi_{k_3}(1) \\ \phi_{k_1}(2) & \phi_{k_2}(2) & \phi_{k_3}(2) \\ \phi_{k_1}(3) & \phi_{k_2}(3) & \phi_{k_3}(3) \end{vmatrix}$$

Through the generalization of the above argument, an eigenfunction capable of describing an N-electron system can be expressed as follows.

$$\Phi(1,2,\ldots,N) = \text{const} \times \begin{vmatrix} \phi_{k_1}(1) & \phi_{k_2}(1) & \ldots & \phi_{k_N}(1) \\ \ldots & \ldots & \ldots & \ldots \\ \ldots & \ldots & \ldots & \ldots \\ \phi_{k_1}(N) & \phi_{k_2}(N) & \ldots & \phi_{k_3}(N) \end{vmatrix} \tag{4.2}$$

The constant in (4.2) is the normalization constant. The state described by the above eigenfunction is called an antisymmetrized state because of the reversal of sign with the exchange of electrons. Fermi particles such as the electron exist only in antisymmetrized states. This fact is called the Pauli principle (W. Pauli, 1924). Needless to say, the principle has nothing to do with the Schrödinger equation.

The value of any determinant is 0 when any two columns or rows coincide. Therefore, when the k_i-th spin orbital ϕ_{k_i} is the same as the k_j-th spin orbital ϕ_{k_j} in the above determinant, the eigenfunction $\Phi(1,2,\ldots,N)$ in (4.2) must be 0, which means that the corresponding state is not allowed. In other words, the Pauli principle states that two electrons cannot occupy one and the same spin orbital $\phi_{k_i} = \phi_{k_j}$. Thus, the principle is alternatively called the Pauli *exclusion* principle. The principle underlies the electron shell model as the building-up principle of multi-electron atoms to be discussed in Sect. 4.2. The exclusion principle plays a crucial role also in considering the electronic structure of molecules in Chaps. 6 through 8, in connection with the concept of the molecular orbital, an analogue of the atomic orbital in multi-electron atoms (cf. Fig. 4.1 and Chaps. 6 through 8).

Since the eigenfunction in (4.2) is a sum of $N!$ products of the spin orbitals, it may be thought that the square of the absolute value of $\Phi(1,2,\ldots,N)$ to represent the probability of finding the N electrons is comprised of a great number of constituents $(N!)^2$. In fact, the number is reduced to $N!$, which is shown below, in the process of determining the normalization constant of the eigenfunction.

The normalization constant is determined by:

$$\int \cdots \int \Phi^*(1,\ldots,N)\Phi(1,\ldots,N)\mathrm{d}\tau_1 \ldots \mathrm{d}\tau_N$$
$$= (\mathrm{const})^2 \int \cdots \int (N! \text{ products})^* \times (N! \text{ products})\mathrm{d}\tau_1 \ldots \mathrm{d}\tau_N = 1 \quad (4.3)$$

In the above equation the symbol $\mathrm{d}\tau_i$ stands for the product of the volume element of the i-th electron in the spatial coordinate $\boldsymbol{r}_i$ and in the spin coordinate σ_i. As for the integration with regard to the spin coordinate, the integral is replaced with simple sums on account of the orthonormality of the spin functions $\alpha(\sigma)$ and $\beta(\sigma)$ as explained in Sect. 3.4.3. Although the spin orbitals $\phi_{k_i}(i) \equiv \phi_{k_i}(\boldsymbol{r}_i)\alpha(\sigma_i)$ and $\phi_{k_i}(i) \equiv \phi_{k_i}(\boldsymbol{r}_i)\beta(\sigma_i)$ are yet unknown, they can be regarded as normalized according to Sect. 2.1.5. Also, they can be assumed to be orthogonal without loosing generality because techniques such as the Schmidt orthogonalization, explained in Sect. 2.1.14, allow them to be orthogonalized at any time.

Equation (4.3) is simplified by the orthonormality of the spin orbitals; the multiple integral, $\int \phi_{k1}^*(1)\phi_{l_1}(1)\mathrm{d}\tau_1 \int \phi_{k_2}^*(2)\phi_{l_2}(2)\mathrm{d}\tau_2 \ldots \int \phi_{k_N}^*(N)\phi_{l_N}(N)\mathrm{d}\tau_N$ becomes 1 only when $k_1 = l_1$, $k_2 = l_2$, ..., $k_N = l_N$ otherwise, it must be 0. Therefore, the integral in the middle of (4.3) turns out to be $N!$, from which the normalization constant is determined as $1/\sqrt{N!}$.

In conclusion, the eigenfunction Φ of N-electron systems is represented by the following Slater determinant as far as the one-electron approximation is valid.

$$\Phi(1,2,\ldots,N) = \frac{1}{\sqrt{N!}} \begin{vmatrix} \phi_{k_1}(1) & \ldots & \phi_{k_N}(1) \\ \ldots & \ldots & \ldots \\ \phi_{k_1}(N) & \ldots & \phi_{k_N}(N) \end{vmatrix}$$
$$\equiv \frac{1}{\sqrt{N!}} |\phi_{k_1}(1)\phi_{k_2}(2)\ldots\phi_{k_N}(N)| \quad (4.4)$$

$$\equiv \|\phi_{k_1}(1)\phi_{k_2}(2)\ldots\phi_{k_N}(N)\| \quad (4.5)$$

The right-hand sides of (4.4) and (4.5) are abbreviated expressions for the determinant.

4.2 The Energy of Multi-Electron Atoms: A Qualitative Consideration and the Concept of Electronic Shell Structure

As stated in the preceding section, the eigenfunction $\Phi(\boldsymbol{r}_1\sigma_1, \boldsymbol{r}_2\sigma_2, \ldots, \boldsymbol{r}_N\sigma_N)$ for the multi-electron system is constructed from the spin orbitals of individual electrons within the one-electron approximation. The expression in terms of the Slater determinant satisfies automatically the Pauli exclusion principle for the electron, which is a Fermi particle.

In this preparatory section a qualitative consideration will be made on the energy of a single electron in the multi-electron atom. The average potential felt by an electron in multi-electron atoms is formally written as $V(1)$, where the coordinate $\boldsymbol{r}_1\sigma_1$ is abbreviated as 1. Then, the eigenfunction $\phi_i(1)$ associated with the energy eigenvalue ε_i should be determined by the Schrödinger equation:

$$\hat{H}\phi_i(1) = \left[-\frac{\hbar^2}{2m}\Delta_1 + V(1)\right]\phi_i(1) = \varepsilon_i\phi_i(1) \tag{4.6}$$

Although the explicit form of $V(1)$ is unknown, it must comprise the contribution from the nucleus, i.e., $-Ze^2/(4\pi\varepsilon_0 r_1)$ and the contributions from all the other electrons, i.e., $+\sum_{j\neq 1}^{N} e^2/(4\pi\varepsilon_0 r_{1j})$. Therefore, the potential $V(1)$ cannot be expressed in a simple form nor can it be spherically symmetric, contrary to the case of one-electron atoms.

However, in order for us to proceed further, it is assumed that the effective contribution from all the other electrons averaged results in spherical symmetry. In so far as this approximation is admitted the eigenfunction $\phi(1)$ is separated into two components $R(r)$ and $Y(\theta,\varphi)$. As for the non-Coulombic potential $V(1)$ ($\equiv V(r)$), it can be sketched qualitatively as in the solid curve in Fig. 4.1. The eigenfunction $\phi_i(1)$ in (4.6) is now rewritten as $\phi(\boldsymbol{r}) = R_{n,l}(r)Y_{l,m}(\theta,\varphi)$ just like the eigenfunction for a one-electron atom in Sect. 3.2.5. Hereafter, this one-electron function will be called the atomic orbital.

Figure 4.1 depicts a conceptual potential to be felt by a single electron in the Li atom. The figure indicates that the potential approaches that for an electron in the field of Li^{3+} at small distances and tends to the potential in the field of H^+ at large distances.

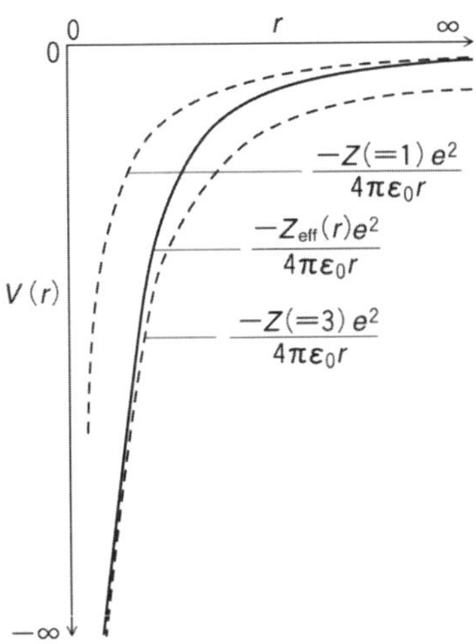

Fig. 4.1. A conceptual sketch of the potential felt by a single electron in the Li atom

In the case of the H atom with a single electron, the total energy is given by (3.17) which depends upon only the principal quantum number n due to the accidental degeneracy (cf. Sect. 3.3.1). Accordingly, the wavelength λ associated with the electronic transition between the states of energies E_m and E_n is determined by the Rydberg formula in (1.12), i.e. $E_m - E_n = hc_0/\lambda = hc_0 R_\infty(1/m^2 - 1/n^2)$.

In the case of the Li atom with three electrons, the photon energy to induce the electronic transition corresponds to the difference of the total energy of the three-electron atom. Therefore, the transition energy cannot be directly related with the energy of a single electron, in contrast to the H atom. Since each electron in the Li atom is assumed to be in a non-Coulombic potential shown in Fig. 4.1, the accidental degeneracy will be removed and the one-electron energy in (4.6) is anticipated to depend on both the principal quantum number n and the azimuthal quantum number l. Reference to Fig. 3.8 suggests that the electron tends to approach the nucleus for a smaller value of l, which should lower the potential energy. In fact, Fig. 4.2 obtained from the experimentally observed spectra shows the expected removal of the degeneracy and the lowering of energy for a smaller value of l among energies with a common value of n.

An energy state resulting from the split of the degeneracy due to the electrostatic interaction among the electrons is called a *term*. For clarity in illustrating the dependence of the splitting upon both n and l, the energy states in Fig. 4.2 are shown in separate columns according to the different l values. Energy diagrams such as in Fig. 4.2 are called Grotrian diagrams after W. Grotrian (1928). At present, the Grotrian diagram has been obtained for almost all atoms and their ions.

The slanted lines in Fig. 4.2 indicate that transitions take place between terms connected by lines. It is seen from the figure that only transitions

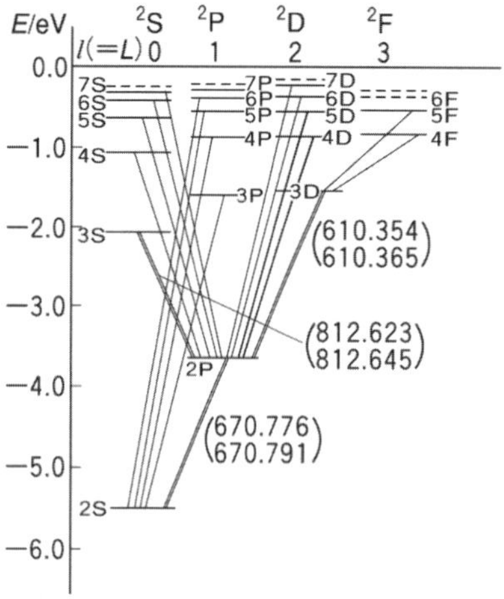

Fig. 4.2. Grotrian diagram for the Li atom

satisfying the relation $|\Delta l| = 1$ are allowed. Such a relation is based on what is called a selection rule* for optical transitions. The symbols S, P, D, … and the dual lines between the terms of 2S and 2P, e.g., will be explained in Sect. 4.3.1. The numbers in parenthesis represent the wavelength of light in units of nm.

Since the potential in Fig. 4.1 is not known exactly, the eigenvalues ε_i in (4.6) for the electrons in the Li atom cannot be determined precisely. However, as far as we regard a single electron as representing the optical transition of the three-electron atom, an approximate one-electron energy corresponding to the above ε_i can be estimated from the observed optical spectra. Figure 4.3 demonstrates such an approximate one-electron energy for the Li atom in comparison with the exact one-electron energy for the H atom.

It is seen from Fig. 4.3 that the one-electron energies of the Li atom to be associated with the atomic orbital $\phi(\boldsymbol{r}) = R_{n,l}(r)Y_{l,m}(\theta, \varphi)$ are divided roughly according to the principal quantum number n. The energy states corresponding to $n = 1, 2, 3, \dots$ are customarily assigned to K, L, M, … shells. The states with the same value of n but different values of l are called subshells or subordinate shells. The term 'electronic shell structure' in the subtitle of this section is intended to describe such an infrastructure of electrons in multi-electron atoms.

It is assumed that the potential in (4.6) is averaged over all directions so that the one-electron eigenfunction $\phi_i(1)$ in the same equation can be

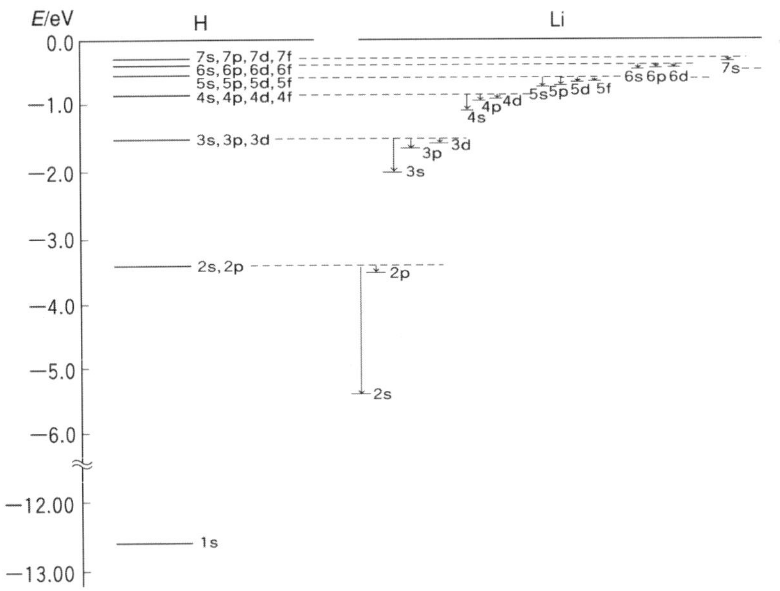

Fig. 4.3. One-electron energies for the H and Li atoms

separated into radial and angular components just like in the case of the H atom. Such an approximation is called the central field approximation. Under this approximation the atomic orbital $\phi(\boldsymbol{r}) = R_{n,l}(r)Y_{l,m}(\theta, \varphi)$ is classified by the azimuthal quantum number l and the magnetic quantum number m, where m assumes $2l + 1$ values of $m = 0, \pm 1, \pm 2, \ldots, \pm l$. The corresponding $2l+1$ atomic orbitals are energetically degenerate in the absence of a magnetic field. As for the electron spin, the two states of $m_s = \pm 1/2$ are also degenerate under the field free condition. Thus, a single subshell can accommodate up to $2(2l + 1)$ electrons.

On the basis of the concept of the electronic shell structure, it is assumed that the electrons in multi-electron atoms in their ground electronic state start the occupation of the shells from those of the lowest energy. Such an assumption is associated with the building-up principle or the Aufbau principle briefly mentioned in Sect. 4.1.

The fifth column of Table 4.1 gives the electronic shell structure or the electronic configuration of multi-electron atoms constructed on the basis of the building-up principle above. This table is consistent in many ways with the periodic table which was constructed by accumulating empirical knowledge long before the birth of quantum mechanics. According to the electronic shell structure in Table 4.1, the outermost shell of He, Ne, and Ar is completely occupied by electrons to give a closed shell. As is well known, the atoms of the rare gas elements are *almost* inert. In these atoms such relations as $|Y_{0,0}|^2 = 1/(4\pi)$ and $|Y_{1,+1}|^2 + |Y_{1,0}|^2 + |Y_{1,-1}|^2 = 3/(4\pi)$ hold good, which implies that the atoms are spherical symmetric as a whole. On the other hand, the outermost shells of atoms with atomic numbers $Z = 1$ and $Z = 3 \sim 9$ and $11 \sim 17$ are not completely occupied by electrons to give open shell electronic structures. Along with the rare gas atoms these atoms are inclusively classified as representative (or main group) elements. As is well known, the chemical properties of the atoms of representative elements belonging to the same family in the periodic table are similar, with the exception of the H atom in the first family.

Table 4.1 lists only the first part of the periodic table and the Ag atom, of atomic number $Z = 47$. The Ag atom is included in the table for the heuristic purpose of understanding the Stern-Gerlach experiment mentioned in Sect. 1.3. From $Z = 21$ (Sc) on, the d-subshell starts being occupied by electrons. The elements having partially occupied d-subshells are called the (outer) transition elements. They are the main subject of inorganic and organometallic complex chemistry and magnetism of matter. This important group of elements, however, is not treated in the present book because we are mainly concerned with the understanding of simple chemical bonds involving electrons in s and p-subshells. The eigenfunction of the d-electrons in one-electron atoms, however, is given in Table 3.1 and Fig. 3.5.

Table 4.1. Electronic shell structure of multi-electron atoms

Z	Element	Period	Family	Shell Structure K L M	m +1	m 0	m −1	M	M_S	Level Symbol
1	H	1	1	$(1s)^1$		↑		0	1/2	$^2S_{1/2}$
2	He	1	18	$(1s)^2$		↑↓		0	0	$^1S_0$
3	Li	2	1	$(1s)^2(2s)^1$		↑		0	1/2	$^2S_{1/2}$
4	Be	2	2	$(1s)^2(2s)^2$		↑↓		0	0	$^1S_0$
5	B	2	13	$(1s)^2(2s)^2(2p)^1$	↑	–	–	1	1/2	$^2P_{1/2}$
6	C	2	14	$(1s)^2(2s)^2(2p)^2$	↑	↑	–	1	1	$^3P_0$
7	N	2	15	$(1s)^2(2s)^2(2p)^3$	↑	↑	↑	0	3/2	$^4S_{3/2}$
8	O	2	16	$(1s)^2(2s)^2(2p)^4$	↑↓	↑	↑	1	1	$^3P_2$
9	F	2	17	$(1s)^2(2s)^2(2p)^5$	↑↓	↑↓	↑	1	1/2	$^2P_{3/2}$
10	Ne	2	18	$(1s)^2(2s)^2(2p)^6$	↑↓	↑↓	↑↓	0	0	$^1S_0$
11	Na	3	1	$(1s)^2(2s)^2(2p)^6(3s)^1$		↑		0	1/2	$^2S_{1/2}$
12	Mg	3	2	$(1s)^2(2s)^2(2p)^6(3s)^2$		↑↓		0	0	$^1S_0$
13	Al	3	13	$(1s)^2(2s)^2(2p)^6(3s)^2(3p)^1$	↑	–	–	1	1/2	$^2P_{1/2}$
14	Si	3	14	$(1s)^2(2s)^2(2p)^6(3s)^2(3p)^2$	↑	↑	–	1	1	$^3P_0$
15	P	3	15	$(1s)^2(2s)^2(2p)^6(3s)^2(3p)^3$	↑	↑	↑	0	3/2	$^4S_{3/2}$
16	S	3	16	$(1s)^2(2s)^2(2p)^6(3s)^2(3p)^4$	↑↓	↑	↑	1	1	$^3P_2$
17	Cl	3	17	$(1s)^2(2s)^2(2p)^6(3s)^2(3p)^5$	↑↓	↑↓	↑	1	1/2	$^2P_{3/2}$
18	Ar	3	18	$(1s)^2(2s)^2(2p)^6(3s)^2(3p)^6$	↑↓	↑↓	↑↓	0	0	$^1S_0$
⋮	⋮	⋮	⋮	⋮	⋮	⋮	⋮			
47	Ag	5	11	$\ldots(4s)^2(4p)^6(4d)^{10}(5s)^1$		↑		0	1/2	$^2S_{1/2}$
⋮	⋮	⋮	⋮	⋮	⋮	⋮	⋮			

4.3 The Angular Momentum of Multi-Electron Atoms: The Total Angular Momentum as a Physical Quantity to Characterize the State of Multi-Electron Atoms

According to the one-electron approximation and the building-up principle stated in the preceding section, multi-electron atoms in their ground electronic state are considered to be constructed by consecutively supplying electrons to the bare nucleus, starting with the atomic orbital of the lowest energy. The electronic shell structure or the electronic configuration in Table 4.1 is based on the above hypothesis. As far as this concept of the building-up is valid, not only the energy but also the orbital angular momentum and the electron spin of multi-electron atoms must be evaluated because each electron is uniquely characterized not only by its energy but also by its two angular momenta $\boldsymbol{l}$ and $\boldsymbol{s}$. In this section the result of this building-up (or coupling) of the constituent angular momenta will be discussed in view of the characterization of multi-electron atoms.

First, we will return to the one-electron atom in Sect. 4.3.1 and think about the coupling which exists between the orbital angular momentum and the electron spin. Then, in Sect. 4.3.2, the composite of these angular momenta combined for all electrons in a multi-electron atom will be considered. As a result, it will be shown that the resultant orbital angular momentum, electron spin, and the total angular momentum are crucial in the characterization of atoms as a whole.

4.3.1 Spin-Orbit Interaction in One-Electron Atoms

In one-electron atoms, such as the H atom, there are only two angular momenta; the orbital angular momentum and the electron spin (note that nuclear spins are ignored in this book). Up to now, the two angular momenta are assumed, as an approximation, to be independent of each other. As a result of this approximation, the eigenfunction of the system denoted as $\phi(\boldsymbol{r}\sigma)$ ($\equiv \phi$) satisfies the following five eigenvalue equations simultaneously: $\hat{H}\phi = E\phi$, $\hat{\boldsymbol{l}}^2\phi = l(l+1)\hbar^2\phi$, $\hat{l}_z\phi = m\hbar\phi$, $\hat{\boldsymbol{s}}^2\phi = s(s+1)\hbar^2\phi$, and $\hat{s}_z\phi = m_s\hbar\phi$. In order to combine the spatial and spin components of the single electron the solution of the Schrödinger equation $\phi(\boldsymbol{r})$ is interpreted as being equally associated with both of the two spin states (cf. Sects. 1.4.2 and 3.4.3).

However, the above assumption is not exactly correct because of the existence of a small interaction between the orbital angular momentum $\boldsymbol{l}$ and the electron spin $\boldsymbol{s}$; since both angular momenta can be viewed as tiny magnets with the respective magnetic moments of $\boldsymbol{\mu} = \gamma\boldsymbol{l}$ and $\boldsymbol{\mu}_s = \gamma_s\boldsymbol{s}$, there must be a magnetic interaction, though it may be small. Here, the symbols γ and γ_s represent the magnetogyric ratio mentioned in Sect. 1.3. This interaction is called spin-orbit interaction, or spin-orbit coupling. Since the two magnets yield their own magnetic fields, the orbital angular momentum $\boldsymbol{l}$ is affected by the field of $\boldsymbol{\mu}_s$. Conversely, the spin $\boldsymbol{s}$ senses the field of $\boldsymbol{\mu}$. As a result, the two angular momenta $\boldsymbol{l}$ and $\boldsymbol{s}$, cause what is known as 'precession', which is a rotating motion of $\boldsymbol{l}$ and $\boldsymbol{s}$ around the axis of the resultant magnetic field. As shown in Fig. 4.4(a), when $\boldsymbol{l}$ and $\boldsymbol{s}$ rotate around the vectorial sum $\boldsymbol{j}$, the two rotate coherently, which is the precession. Since the direction of $\boldsymbol{j}$ remains fixed permanently, it is anticipated that this vector is qualified to characterize the system. In fact, such stationary physical quantities play an important role in both classical and quantum mechanics so that they are named generally constants of motion. The vector $\boldsymbol{j}$ is called the total angular momentum.

As stated in Sect. 3.4.3, both $\boldsymbol{l}$ and $\boldsymbol{s}$ are representatives of the generalized angular momentum. When two (generalized) angular momenta are added, only some of the resultants are acceptable as the total (generalized) angular momentum. According to the theory of angular momentum, the possible angular momentum functions for an added angular momentum are obtained by a linear combination of simple products of functions for angular momenta to be added through the use of Clebsch-Gordan coefficients*, which are alternatively called Wigner coefficients*, or vector addition coefficients*.

In order to generalize the explanation on the addition of the angular momenta $\boldsymbol{l}$, $\boldsymbol{s}$, and $\boldsymbol{j}$ in Fig. 4.4(a) they are first rewritten as $\boldsymbol{j}_1$, $\boldsymbol{j}_2$, and $\boldsymbol{j}_{12}$, respectively, to remind us that they are all generalized angular momenta. As given in (3.31) in Sect. 3.4.1, the values of the constituents are fixed as $\boldsymbol{j}_1 = \sqrt{j_1(j_1+1)}\hbar$ and $\boldsymbol{j}_2 = \sqrt{j_2(j_2+1)}\hbar$, where j_1 and j_2 are the respective quantum numbers. Similarly, according to (3.32), the z-axis components of $\boldsymbol{j}_1$ and $\boldsymbol{j}_2$, i.e., j_{z1} and j_{z2} are given as $j_{z1} = m_1\hbar$ and $j_{z2} = m_2\hbar$, where the quantum numbers m_1 and m_2 assume the values of $m_1 = j_1, j_1-1, \ldots, -(j_1-1), -j_1$ and $m_2 = j_2, j_2-1, \cdots -(j_2-1), -j_2$. As for the added angular momentum $\boldsymbol{j} \equiv \boldsymbol{j}_{12} = \boldsymbol{j}_1 + \boldsymbol{j}_2$, its z-axis component is given by the scalar sum of the constituents as $j_z \equiv j_{z12} = m_1\hbar + m_2\hbar \equiv m_{12}\hbar = m_j\hbar$. Thus, the quantum number m_{12} runs over the range defined by the maximum (j_1+j_2) to the minimum $-(j_1+j_2)$, with many values of m_{12} occurring more than once.

The fact that the value of $m_j = m_{12}$ varies as above means that the maximum value of the quantum number $j \equiv j_{12}$ for the added angular momentum $\boldsymbol{j} \equiv \boldsymbol{j}_{12}$ must be equal to $j_1 + j_2$. As for the minimum of $j \equiv j_{12}$, it is found to be equal to $|j_1 - j_2|$, as confirmed by substituting the possible values of m_1

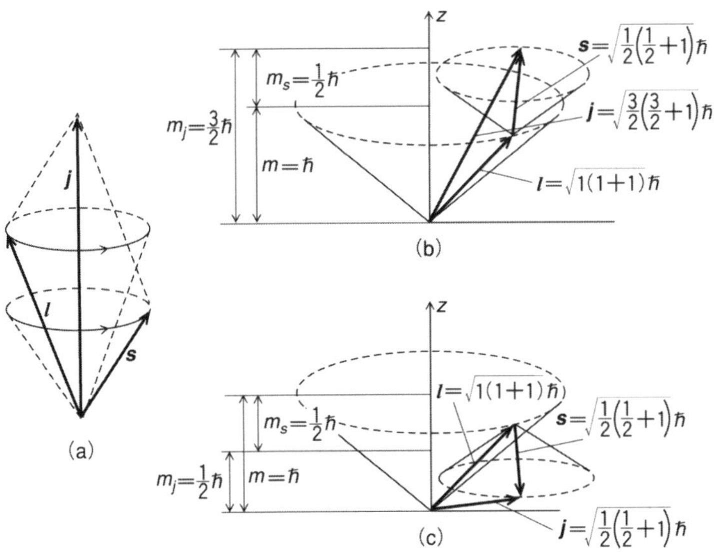

Fig. 4.4. (**a**) The orbital angular momentum $\boldsymbol{l}$ and the electron spin $\boldsymbol{s}$ add vectorially to yield the total angular momentum $\boldsymbol{j}$ of a single electron. While $\boldsymbol{l}$ and $\boldsymbol{s}$ rotate around $\boldsymbol{j}$, the latter vector remains unchanged pointing in a fixed direction. (**b**), (**c**) Two possible resultants for $\boldsymbol{j}$ are obtained from the vector $\boldsymbol{l}$ with $l = 1$, and another vector $\boldsymbol{s}$ with $s = 1/2$. The three vectors satisfy the triangle relation while the three quantum numbers m_l, m_s, and m_i couple as scalars

and m_2 enumerated above into the relationship $m_j \equiv m_{12} = m_1 + m_2$. Thus, we can conclude that the possible quantum numbers of the added angular momentum are restricted to $j_{12} = j_1 + j_2, (j_1 + j_2) - 1, (j_1 + j_2) - 2, \ldots, |j_1 - j_2|$ for $j \equiv j_{12}$, and as $m_j \equiv m_{12} = j_{12}, j_{12} - 1, j_{12} - 2, \ldots, -j_{12}$ for each j_{12}.

As an example of the addition of $\boldsymbol{j}$ from $\boldsymbol{l}$ and $\boldsymbol{s}$, the case of $l = j_1 = 1$ and $s = j_2 = 1/2$ is illustrated in Figs. 4.4(b) and (c). The cones associated with the constituent vectors $\boldsymbol{l}$ and $\boldsymbol{s}$ are projected onto the z-axis and the two scalars m and m_s give the possible quantum number for the added vector $\boldsymbol{j}$ by the relation $m + m_s = m_1 + m_2 = m_{12} = m_j$. It is said that the three vectors $\boldsymbol{l}$, $\boldsymbol{s}$, and $\boldsymbol{j}$ satisfy the triangle relationship.

The above explanation of the spin-orbit interaction is qualitative in the sense that an analogy of the interaction between two magnets has been drawn. More properly, the spin-orbit interaction is represented as a perturbing Hamiltonian $\hat{H}'$, which is added to the unperturbed Hamiltonian and then, the eigenvalue equation for energy is solved. In the absence of the perturbation the eigenfunction of the system satisfies simultaneously the following five eigenvalue equations: $\hat{H}\phi = E\phi$, $\hat{\boldsymbol{l}}^2\phi = l(l+1)\hbar^2\phi$, $\hat{l}_z\phi = m\hbar\phi$, $\hat{\boldsymbol{s}}^2\phi = s(s+1)\hbar^2\phi$, and $\hat{s}_z\phi = m_s\hbar\phi$ as stated above. Under such a situation the quantum numbers l, m, s, and m_s are called good quantum numbers. However, in the presence of the perturbation, the above five are replaced with the following three equations: $(\hat{H} + \hat{H}')\phi = E\phi$, $\hat{\boldsymbol{j}}^2\phi = j(j+1)\hbar^2\phi$, and $\hat{j}_z\phi = m_j\hbar\phi$, and the good quantum numbers become j and m_j.

The explicit form of the spin-orbit interaction can, to a good approximation, be expressed as $\hat{H}' = A\hat{\boldsymbol{l}} \cdot \hat{\boldsymbol{s}}$, where A is a proportionality constant. The energy change due to this perturbation is estimated by a first order approximation as follows, where ϕ and ϕ^* are the eigenfunctions in the absence of the perturbation:

$$\begin{aligned} \Delta E &= \int \phi^* A\hat{\boldsymbol{l}} \cdot \hat{\boldsymbol{s}}\phi \mathrm{d}v = \int \phi^*(A/2)(\hat{\boldsymbol{j}}^2 - \hat{\boldsymbol{l}}^2 - \hat{\boldsymbol{s}}^2)\phi \mathrm{d}v \\ &= (A/2)\left[j(j+1) - l(l+1) - s(s+1)\right]\hbar^2 \end{aligned} \tag{4.7}$$

In (4.7), the relation $\hat{\boldsymbol{j}}^2 = \hat{\boldsymbol{j}} \cdot \hat{\boldsymbol{j}} = (\hat{\boldsymbol{l}} + \hat{\boldsymbol{s}}) \cdot (\hat{\boldsymbol{l}} + \hat{\boldsymbol{s}}) = \hat{\boldsymbol{l}}^2 + \hat{\boldsymbol{s}}^2 + 2\hat{\boldsymbol{l}} \cdot \hat{\boldsymbol{s}}$ is employed. Also, note that the equality $\hat{\boldsymbol{l}} \cdot \hat{\boldsymbol{s}} = \hat{\boldsymbol{s}} \cdot \hat{\boldsymbol{l}}$ holds because $\hat{\boldsymbol{l}}$ and $\hat{\boldsymbol{s}}$ are commutative.

In the case of Figs. 4.4(b) and (c) the quantum number j, running from $j = l + s$, $(l + s) - 1, \ldots, |l - s|$ is restricted to $j = 3/2$ and $j = 1/2$. Thus, the energy state is split into two, separated by an amount ΔE in (4.7). If the system is in a magnetic field, the state with quantum number j is further split into $2j + 1$ states, with the magnetic quantum number being $m_j = j, j-1, \ldots, -(j-1), -j$. This is nothing more than the Zeeman effect in the presence of the spin-orbit interaction in one-electron atoms.

Within the non-relativistic limit, the absorption line of the H atom associated with the transition from the 1s to the 2p state should be single, of a wavelength given by substituting $m = 1$ and $n = 2$ in (1.12) in Sect. 1.4.1.

However, if the spin-orbit interaction is taken into account, the single line splits into a doublet, although the separation between the doublet is only $0.4\,\mathrm{cm}^{-1}$ as compared with an energy of $82{,}259\,\mathrm{cm}^{-1}$ for the 1s→2p transition.

Similar splittings are observed for the Li atom in Fig. 4.2. In the Li atom, the electron in the outermost 2s shell is optically excited to np shells, which is similar to the 1s→2p transition in the H atom. An energy state arising after splitting by the spin-orbit interaction is called a *level*. This is to be differentiated from the *term* in Sect. 4.2, which is associated with the energy states split simply through the electrostatic interaction. In the absence of magnetic fields each level is $(2j+1)$-fold degenerate.

Since the energy separation among the levels belonging to the same term is very small in the above cited examples, the spin-orbit interaction may appear insignificant in chemical bonding. However, in multi-electron atoms both $\boldsymbol{l}_i$ and $\boldsymbol{s}_i$ must be coupled, as will be argued in Sects. 4.3.2 and 4.3.3 below. As a result, the electrostatic interaction varies from one term to another. This difference between terms causes significant effects in the formation of chemical bonds. That is to say, the influence of the electron spin in multi-electron atoms is indirect but significant in the formation of chemical bonds, even though the spin-orbit interaction in one-electron atoms of small atomic numbers is negligibly small.

4.3.2 Couplings of Orbital Angular Momentum and Spin in Multi–Electron Atoms: Russell-Saunders Coupling

In the preceding section the perturbation Hamiltonian due to the spin-orbit interaction in one-electron atoms was given, without proof, as $\hat{H}' = A\hat{\boldsymbol{l}} \cdot \hat{\boldsymbol{s}}$. It is known that the constant A is roughly proportional to the atomic number Z to the power of 4. Therefore, in relatively light atoms, let's say, those belonging to the first to the third period of the periodic table as listed in Table 4.1, the spin-orbit interaction is small compared with heavier atoms. This means that $\boldsymbol{l}$ and $\boldsymbol{s}$ do not interact strongly with each other.

Instead, in these lighter atoms, the coupling among $\boldsymbol{l}_i$'s, and that among $\boldsymbol{s}_i$'s, becomes more significant. It is interpreted that all the $\boldsymbol{l}_i$'s are coupled to yield the overall orbital angular momentum of $\boldsymbol{L}$. Likewise, all the $\boldsymbol{s}_i$'s are regarded as constructing the overall electron spin $\boldsymbol{S}$. It is then considered that $\boldsymbol{L}$ and $\boldsymbol{S}$ are coupled to result in the final total angular momentum $\boldsymbol{J} = \boldsymbol{L} + \boldsymbol{S}$, which is a constant of motion for these multi-electron atoms. This mode of coupling is called Russell-Saunders coupling or $\boldsymbol{LS}$ coupling.

In contrast to the light atoms considered above, the spin-orbit coupling of each electron becomes larger in heavier atoms. In these atoms, $\boldsymbol{l}_i$ and $\boldsymbol{s}_i$ of each electron are more appropriately coupled to give $\boldsymbol{j}_i$ first, just as in one-electron atoms discussed in Sect. 4.3.1, and then all the $\boldsymbol{j}_i$'s are coupled to yield the overall angular momentum $\boldsymbol{J}$, which is again a constant of motion for these multi-electron atoms. This coupling scheme is called $\boldsymbol{jj}$ coupling.

The two modes of coupling themselves are merely a matter of preference in the process of coupling. What is physically meaningful is the final total angular momentum $\boldsymbol{J}$, which is a constant of motion of the system, be it for lighter or heavier atoms. Since we are concerned with the chemical bonding between relatively light atoms, for which the concept of $\boldsymbol{LS}$ coupling is easier to conceptualize, $\boldsymbol{jj}$ coupling will not be subject to further discussion in this book. For the case of $\boldsymbol{LS}$ coupling we will consider the usefulness of the interim coupling among $\boldsymbol{l}_i$'s and among $\boldsymbol{s}_i$'s. This is because both $\boldsymbol{L}$ and $\boldsymbol{S}$, though not real constants of the motion, are convenient devices in differentiating the energy states of multi-electron atoms, as will be shown below.

First, allow me to remind the reader that the angular momenta $\boldsymbol{l}_i$ and $\boldsymbol{s}_i$ of a single electron are given as $\sqrt{l_i(l_i+1)}\hbar$ and $\sqrt{s_i(s_i+1)}\hbar$, and their z-axis components as $m_i\hbar$ ($m_i = 0, \pm 1, \pm 2, \ldots, \pm l$) and $m_{si}\hbar$ ($m_{si} = \pm 1/2$).

In $\boldsymbol{LS}$ coupling, the quantum numbers L and S, which determine the coupled angular momenta $\boldsymbol{L}$ and $\boldsymbol{S}$, are obtained by the following procedure.

In the preceding Sect. 4.3.1, the procedure to determine the quantum number j ($\equiv j_{12}$) for the total angular momentum $\boldsymbol{j}_{12}$ from $\boldsymbol{l}$ ($\equiv \boldsymbol{j}_1$) and $\boldsymbol{s}$ ($\equiv \boldsymbol{j}_2$) has been described for one-electron atoms. This procedure can be extended to $\boldsymbol{LS}$ coupling in multi-electron atoms. First, we consider the coupling among $\boldsymbol{l}_i$'s starting with $\boldsymbol{l}_1 \equiv \boldsymbol{j}_1$ and $\boldsymbol{l}_2 \equiv \boldsymbol{j}_2$, which are coupled as $\boldsymbol{j}_{12} = \boldsymbol{j}_1 + \boldsymbol{j}_2$. Next, the resulting $\boldsymbol{j}_{12}$ is coupled with $\boldsymbol{l}_3 \equiv \boldsymbol{j}_3$ of the third electron to obtain $\boldsymbol{j}_{12} + \boldsymbol{j}_3 \equiv \boldsymbol{j}_{123}$. Continuing the same procedure, the fourth electron is coupled to obtain $\boldsymbol{j}_{1234}$ from $\boldsymbol{j}_{123}$ and $\boldsymbol{j}_4$.

As a result of these successive couplings of $\boldsymbol{l}_i$ of k electrons, the quantum number L for the total orbital angular momentum $\boldsymbol{L}$ turns out to be $L = (l_1 + l_2 + \cdots + l_k), (l_1 + l_2 + \cdots + l_k) - 1, (l_1 + l_2 + \cdots + l_k) - 2, \ldots, \geq 0$. The coupling of $\boldsymbol{s}_i$ of k electrons to $\boldsymbol{S}$ is similar to the above except that the acceptable quantum number S has a minimum of 0 or 1/2 depending upon whether k is even or odd, respectively, because the quantum number s_i of a single electron is 1/2. To be specific, when k is even, the allowed values for S are $S = (s_1 + s_2 + \cdots + s_k), (s_1 + s_2 + \cdots + s_k) - 1, (s_1 + s_2 + \cdots + s_k) - 2, \ldots, 0 = k/2, k/2 - 1, k/2 - 2, \ldots, 0$. Likewise, when k is odd, $S = k/2, k/2 - 1, k/2 - 2, \ldots, 1/2$.

As for the z-axis components of $\boldsymbol{L}$ and $\boldsymbol{S}$, the quantum numbers M and M_S in $L_z = M\hbar$ and $S_z = M_S\hbar$ are given by the scalar sums of $M = \sum m_i$ and $M_S = \sum m_{si}$, both of which are restricted to $M = 0, \pm 1, \pm 2, \ldots, \pm L$ and $M_S = +S, +(S-1), +(S-2), \ldots, -(S-2), -(S-1), -S$. These $2L+1$ and $2S+1$ states are degenerate in the absence of external magnetic fields.

The final total angular momentum $\boldsymbol{J}$ is obtained by the coupling of $\boldsymbol{L}$ and $\boldsymbol{S}$ as $\sqrt{J(J+1)}\hbar$, the quantum number J being restricted to $J = (L + S), (L+S) - 1, (L+S) - 2, \ldots, |L - S|$. The z-axis components of $\boldsymbol{J}$ are given by $J_z = M_J\hbar$ where the quantum number M_J is restricted to $M_J = +J, +(J-1), (J-2), \ldots, -(J-2), -(J-1), -J$. These $2J+1$ states are also degenerate in the absence of magnetic fields. An example of this degeneracy will be given in Sect. 8.2.3.

From the above, the electronic states of multi-electron atoms can be designated by the three quantum numbers L, S, and J. A set of these three is concisely denoted as ${}^{2S+1}L_J$, which corresponds to a level defined in Sect. 4.3.1 above. The superscript $2S+1$ is called the spin multiplicity. When the value of L is 0, 1, 2, 3, ..., the state is denoted by the letters S, P, D, F, ..., respectively. This convention originates from the convention of spectroscopy to symbolize the states $l = 0, 1, 2, 3, \ldots$ of one-electron atoms by the letters s, p, d, f, ... The ninth column of Table 4.1 shows an example of the use of these designations to represent the levels of the atoms in their ground electronic state. In the H atom, with a single electron, the symbol becomes simply ${}^2\mathrm{S}_{1/2}$. For the atoms with a closed electronic shell both L and S must be 0 because of the relations $M = \sum m_i = 0$ and $M_S = \sum m_{si} = 0$ so that the symbol of the level becomes uniquely ${}^1\mathrm{S}_0$. Accordingly, it is the electronic configuration of incompletely filled electronic shells that determines level symbols other than ${}^1\mathrm{S}_0$. For example, in the Li atom in its ground electronic state, the level is determined only by the outermost 2s electron to be ${}^2\mathrm{S}_{1/2}$, which is the same as that for the H atom. The two lines for the Li atom with the wavelengths of 670.776 nm and 670.791 nm in Fig. 4.2 correspond to the transitions from the level of the ground state to the two levels ${}^2\mathrm{P}_{1/2}$ and ${}^2\mathrm{P}_{3/2}$ due to the optically excited 2p electron.

The fifth through ninth columns in Table 4.1 indicate the electronic shell structure of each atom in its ground electronic state. As for the illustrations in the sixth column and the assignments in the seventh and eighth columns, a critical comment will be made in Sect. 4.3.3 later. For the moment, I will give a rather naive explanation of the illustrations by giving as example the case of the C atom. According to the illustration, the two outermost electrons of the up-spin are to occupy the orbitals of $m = +1$ and $m = 0$. Reference to Figs. 3.6 and 3.7 indicates that the directional distribution of these electrons are different so that the repulsive interaction between them would be smaller than in the case when the two electrons were confined in one of the three orbitals $m = +1$, or $m = 0$, or $m = -1$ with their spins being forced to be anti-parallel by the Pauli exclusion principle. In the latter case the repulsive interaction between the two electrons would certainly be higher than the case shown in the illustration.

4.3.3 Coupled Angular Momentum of Multi-Electron Atoms (Case 1): Concrete Examples of the Procedure of Coupling

In this section two concrete examples of the coupling procedure for determining angular momentum will be shown for the ***LS*** coupling case. As the first example, consider the coupling for two electrons occupying different subshells of $n \neq n'$ but $l = l' = 1$. Since the two are in different orbitals np and

n'p, there is no restriction due to the Pauli exclusion principle as to the distribution of the electrons to the orbitals. The possible quantum numbers for the coupled angular momentum are limited to $L = 2, 1, 0$ and $S = 1, 0$ from the argument in the preceding Sect. 4.3.2. Thus, the possible levels are $^3\mathrm{D}_{3,2,1}$, $^1\mathrm{D}_2$, $^3\mathrm{P}_{2,1,0}$, $^1\mathrm{P}_1$, $^3\mathrm{S}_1$, and $^1\mathrm{S}_0$, and, according to the degeneracy of $(2L+1)(2S+1)$, the number of states in each term is calculated to be 15, 5, 9, 3, 3, and 1 totaling to 36 states. The same total number is obtained also by summing up all the $(2J+1)$-fold degenerate levels, i.e., from $^3\mathrm{D}_{3,2,1}$, we have $7+5+3 = 15$ states; from $^1\mathrm{D}_2$, 5 states; from $^3\mathrm{P}_{2,1,0}$, $5+3+1 = 9$ states; from $^1\mathrm{P}_1$, 3 states; from $^3\mathrm{S}_1$, 3 states; and finally from $^1\mathrm{S}_0$, 1 state. In a similar manner, it can be derived immediately that the case $ns \neq n'$s yields ^{1}S and ^{3}S and the case ns and n'p gives ^1P and ^{3}P.

The second example is appropriate to the C atom where $n_1 = n_2 = 2$ and $l_1 = l_2 = 1$, so that the electronic configuration is written as $(2\mathrm{p})^2$. In this case the Pauli exclusion principle reduces the total number of states from 36 to 15. Several techniques to select these 15 states have been proposed (cf. e.g., Steinfeld in Refs. for Chap. 4). Here, I will present the a method by referring to Table 4.2.

The first column of Table 4.2 is for the entry of the magnetic quantum numbers m_1 and m_2 of electrons 1 and 2 and of the sums of $M = m_1 + m_2$ and $M_S = m_{s1} + m_{s2}$, whereas the first row is for the entry of the fifteen possible patterns of distribution of two electrons among the three values of m and two values of m_s (up- and down-spins). The distribution patterns [1] through [15] correspond to the Slater determinant expressions (cf. (4.5)) shown in the last row. The abbreviated expression $(1, \bar{1})$, e.g., stands for the determinant $\|\phi_{m_1=+1}(1)\alpha(1)\phi_{m_2=+1}(2)\beta(2)\|$ where electron 1 is placed into the spatial orbital $\phi_{m_1}(1)$ and the spin function $\alpha(1)$, while electron 2 is into $\phi_{m_2}(2)$ and $\beta(2)$.

Since the maximum value of M in Table 4.2 is 2, the corresponding distribution pattern [1] must be associated with $L = 2$, the term symbol of which is D. Since the value M_S of this term is 0, the spin quantum number

Table 4.2. Fifteen electron distribution patterns compatible with the Pauli exclusion principle for the p^2 electronic configuration

	[1]	[2]	[3]	[4]	[5]	[6]	[7]	[8]	[9]	[10]	[11]	[12]	[13]	[14]	[15]
+1	⇅	↑	↑	↓	↓	↑	↑	↓	↓	–	–	–	–	–	–
0	–	↑	↓	↑	↓	–	–	–	–	⇅	↑	↑	↓	↓	–
−1	–	–	–	–	–	↑	↓	↑	↓	–	↑	↓	↑	↓	⇅
M	2	1	1	1	1	0	0	0	0	0	−1	−1	−1	−1	−2
M_S	0	1	0	0	−1	1	0	0	−1	0	1	0	0	−1	0
	$(1,\bar{1})$	$(1,0)$	$(1,\bar{0})$	$(\bar{1},0)$	$(\bar{1},\bar{0})$	$(1,-1)$	$(1,-\bar{1})$	$(\bar{1},-1)$	$(\bar{1},-\bar{1})$	$(0,\bar{0})$	$(0,-1)$	$(0,-\bar{1})$	$(\bar{0},-1)$	$(\bar{0},-\bar{1})$	$(-1,-\bar{1})$

S is 0, so that distribution pattern [1] is identified with ^{1}D. The following five sets of M and M_S, i.e., $(M = \pm 2, M_S = 0)$, $(M = \pm 1, M_s = 0)$, and $(M = 0, M_s = 0)$ are associated with ^{1}D.

Next, we look for patterns with $M_S = 1$ to find [2], [6], and [11]. Thus, these three must be ascribed to the spin quantum number $S = 1$. Since the value of M of these three are 1, 0, and −1, respectively, the following nine sets of M and M_S, that is to say, $(M = 1, M_S = \pm 1)$, $(M = 1, M_S = 0)$, $(M = 0, M_S = \pm 1)$, $(M = 0, M_S = 0)$, $(M = -1, M_S = \pm 1)$, and $(M = -1, M_S = 0)$ are associated with ^{3}P. By now we have counted fourteen sets of M and M_S as belonging to ^{1}D and ^{3}P. The remaining one set of $(M = 0, M_S = 0)$ is obviously associated with ^{1}S.

In the above procedure, the fifteen distribution patterns are associated with three terms ^{1}D, ^{3}P, and ^{1}S by taking note of the maximum values of M and M_S. Obviously, the terms derivable from distribution patterns having N electrons in the same subshell are the same as the terms obtained from patterns lacking N electrons relative to the completely filled subshell. For example, the O atom with the $2p^4$ electronic configuration is a counterpart of the C atom with the $2p^2$ configuration so that in the O atom also the three terms of ^{1}D, ^{3}P, and ^{1}S obtained for the C atom are derived. Likewise, the pair of $2p^1$ and $2p^5$ yield a common single term, ^{2}P. A somewhat complicated case is the N atom with the $2p^3$ configuration. In this case, the three terms ^{1}D, ^{3}P, and ^{1}S appropriate for the $2p^2$ configuration are combined with the third 2p electron to yield the three terms of ^{2}D, ^{2}P, and ^{4}S. Other examples of some complexity are configurations such as sp^3, where not all the electrons are equivalent. In this case, the first step is to look for the possible terms obtained by combining the equivalent electrons (i.e., three p-electrons) and then the resultant is combined with the non-equivalent electron (i.e., one s-electron). In the case of sp^3 we have ^{2}D, ^{2}P, and ^{4}S from p^3 which are coupled with ^{2}S due to the one s-electron, to result in the six terms of ^{3}D, ^{1}D, ^{3}P, ^{1}P, ^{5}S, and ^{3}S.

The procedure described above is admittedly laborious and tedious. However, the terms to be derived from the major configurations have already been worked out and are listed in textbooks such as the one by Condon and Shortley (cf. Refs. for Chap. 3). For the understanding of chemical bonding, the electronic structure of C, N, and O atoms are important. These are covered in the above explanations.

At this stage, the naive explanation in the last paragraph of the preceding Sect. 4.3.2 will be reexamined. According to Table 4.1, the level symbol of $^3P_0$ is given to the C atom in the ground electronic state and the outermost shell structure is illustrated in the sixth column of that table, which coincides with distribution pattern [2] of Table 4.2. In the following this illustration will be critically discussed.

In multi-electron atoms, the interelectronic Coulombic interaction is overwhelming compared with the subtle magnetic interactions between the orbital angular momentum and the electron spin. Were it not for the Coulombic interaction, the fifteen distribution patterns in Table 4.2 would be virtually degenerate in energy. Owing to this interaction, however, the degeneracy is lifted and the energy eigenvalues as well as the eigenfunctions are split according to the theory of perturbation for a degenerate system (cf. Sect. 2.1.14 and Fig. 4.5). When $\boldsymbol{LS}$ coupling is appropriate to describe the system, the eigenfunction Φ should satisfy simultaneously the eigenvalue equations with regard to the total orbital angular momentum operators and to the total electron spin operators as follows (cf. Sect. 2.1.12).

$$\hat{H}\Phi = E\Phi, \quad \hat{\boldsymbol{L}}^2\Phi = L(L+1)\hbar^2\Phi, \quad \hat{\boldsymbol{S}}^2\Phi = S(S+1)\hbar^2\Phi,$$
$$\hat{L}_z\Phi = M\hbar\Phi, \quad \hat{S}_z\Phi = M_S\hbar\Phi \tag{4.8}$$

The above equations show that the state of multi-electron atoms can be characterized not only by the energy but also by the quantum numbers L, M, S, and M_S. Table 4.2 gives only M and M_S for the fifteen distribution patterns, giving no classification by the quantum numbers L and S. As is seen from Table 4.2, some patterns share the same common eigenvalues of M and M_S, i.e., the pair of [3] and [4] with $M = 1$, $M_S = 0$, the set of [7], [8], and [10] with $M = 0$, $M_S = 0$, and the pair of [12] and [13] with $M = -1$, $M_S = 0$. These sets represent simultaneous eigenfunctions of $\hat{L}_z$ and $\hat{S}_z$ with the common eigenvalues, but they are not necessarily the eigenfunctions of $\hat{\boldsymbol{L}}^2$ and $\hat{\boldsymbol{S}}^2$. In order to obtain such eigenfunctions the degenerate distribution patterns must be linearly combined to satisfy the eigenvalue equations for $\hat{\boldsymbol{L}}^2$ and $\hat{\boldsymbol{S}}^2$. An analogous procedure to get such linear combinations will be demonstrated in a problem of the H_2 molecule later in Sect. 7.2.2. At the moment, I simply remark that the illustrations for the C atom in the sixth through eighth columns of Table 4.1 represent just one of the possible nine states in the term of ^{3}P, which are to be described in general by linear combinations of the degenerate distribution patterns illustrated in Table 4.2.

The subscript 0 in the level symbol of $^3\mathrm{P}_0$ in the last column of Table 4.1 is given because the level with $J = 0$ is ensured to be the most stable after the spin-orbit interaction is taken into account quantitatively (cf. (4.7) in Sect. 4.3.1 and Fig. 4.5).

4.3.4 Coupled Angular Momentum of Multi-Electron Atoms (Case 2): Hund's Rules

In Sect. 4.3.3, the procedure to select possible terms compatible with the Pauli exclusion principle was explained. Since the difference in the spatial distribution of electrons in each term results in different interactions among the electrons, the energy differs among the terms. It is important to know this energy difference and thus the sequential order of the energies of the terms. For this purpose, it is required to compute the one-center Coulomb repulsion integrals F and the Coulomb exchange integrals G, which are called the

Slater-Condon parameters*. Fortunately, these parameters have already been given in textbooks of atomic physics. In the discussion of chemical bonding, the numerical values of the parameters themselves are seldom needed. More useful guidelines exist to know the sequential order of energy, or the stability among various terms. These are Hund's rules, which have been derived empirically from the analysis of a vast amount of atomic spectra (F. Hund, 1925). The three rules consist of the following:

(1) Among the terms derived from the same electronic configuration, the one with the maximum spin multiplicity $2S + 1$ is the most stable, and the stability decreases with the multiplicity.

(2) Among the terms of the same spin multiplicity, the one with the maximum quantum number L for the orbital angular momentum is the most stable.

(3) Terms with the same values for L and S are divided into the following two groups. In atoms where the outermost subshell is filled by electrons up to half the maximum number of electrons allowable for that shell, the level with the minimum quantum number $J = |L - S|$ for the total angular momentum is the most stable. In contrast, in atoms having electrons occupying more than half the subshell, the level with the maximum $J = L + S$ is the most stable.

Hund's rules hold very well for the ground electronic configuration, but not so well for excited electronic configurations. The rules are effective in determining the most stable term within a given electronic configuration, but they are not necessarily reliable when used to predict the order of energy for terms other than the most stable term.

The above statements can be confirmed by reference to Fig. 4.5. For the ground electronic configurations, the order of the terms is consistent with Hund's rules as C ($^3P < {}^1D < {}^1S$), N ($^4S < {}^2D < {}^2P$), and O ($^3P < {}^1D < {}^1S$). However, for the excited electronic configuration $2s2p^3$ of the C atom, the observed order of $^5S < {}^3D < {}^3P < {}^1D < {}^3S < {}^1P$ is not in conformity with Hund's rules, as predicted above.

It was traditionally believed that the physical basis for Hund's first rule was a consequence of the Pauli exclusion principle that the electrons in the triplet state ($S = 1$) were kept apart to lower the interelectronic repulsion energy relative to a corresponding singlet. However, advances in computational accuracy have revealed that this is a misconception naively deduced from the inference that the expectation value of the interelectronic distance would be larger in the triplet. Rigorous calculations indicate that the electrons in a triplet state have a considerably higher probability of being found at intermediate interelectronic separations and an accompanying higher Coulomb repulsion energy than the counterpart singlet ($S = 0$). In reality, the energies of the triplets are lower and Hund's first rule holds because the larger electron-electron repulsion of the triplet is accompanied by an electron-nuclear attraction that more than compensates for the electron-electron repulsion.

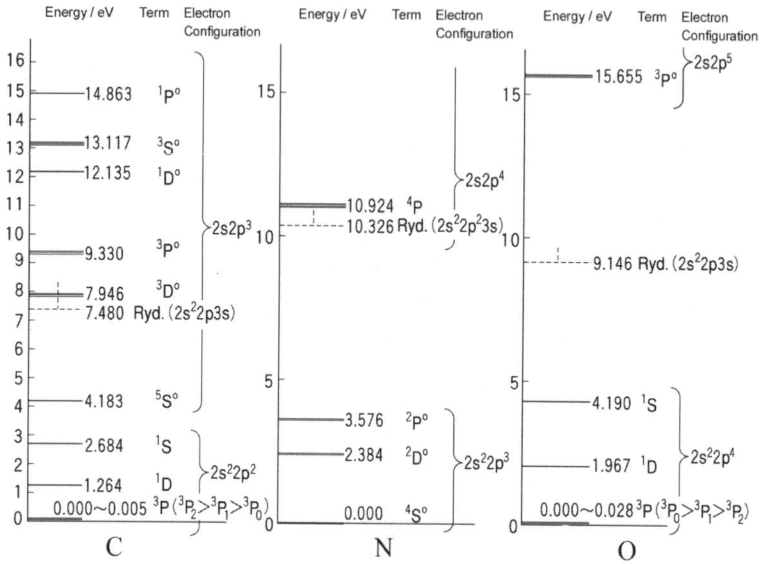

Fig. 4.5. Energy terms of the C, N, and O atoms in their ground and low lying excited electronic configurations. The broken lines indicate Rydberg states which have little relation to chemical bonding and are ignorable. The superscript o attached to the term symbol stands for the 'odd' symmetry with regard to the inversion of electrons in the atoms

4.3.5 Electronic Configuration in Atoms and Valence States

As shown in Tables 4.1, 4.2 and Fig. 4.5, the atoms C, N, and O in their lowest energy state have open shell structures of $2p^2$, $2p^3$, and $2p^4$, respectively. Accordingly, it might be anticipated that these atoms would behave as a divalent, a trivalent, and a tetravalent atom, respectively. In actual chemical bonding, however, the C atom behaves as a tetravalent atom in most cases. This fact suggests that the atomic state in chemical bonds is different from the energy states for the free atom in Fig. 4.5.

To account for this discrepancy Van Vleck introduced the concept of valence state (J.H. Van Vleck, 1933). He conjectured that for the C atom to behave as a tetravalent atom the electronic configuration must be modified from $2s^22p^2$ to $2s2p^3$, in which the four electrons could be distributed over the four atomic orbitals allowing formation of four chemical bonds. In this new configuration the following six terms $^5S^\circ, {}^3D^\circ, {}^3P^\circ, {}^1D^\circ, {}^3S^\circ$, and $^1P^\circ$ are derivable according to the procedure in Sect. 4.3.2. The left column of Fig. 4.5 shows these terms, in which both the orbital angular momentum and the electron spin are given definite values of L and S. For example, in the term $^5S^\circ$ all four electrons are associated with the orbital angular momentum 0, and the electron spins are either all up or all down. The atom as such, however, cannot participate in the formation of CH_4, e.g., because the four electron spins of the C atom participating in the C-H bond formation should have no correlation among them, i.e., the belief that all the four spins are up or down must be removed. Such a randomized spin state is a virtual state to be constructed by a suitable linear combination of the above six real terms. This virtual state is named the valence state, which is an average of the six terms ranging from the term $^5S^\circ$ at the minimum energy to the term $^1P^\circ$ at the maximum energy (cf. Fig. 4.5, left). The average energy was found to be about 8.26 eV by a complicated process of averaging to obtain uncorrelated spin states. Once the valence state is determined, one can imagine that the free C atom in the ground electronic state is excited to this virtual valence state, in which the C atom can attract smoothly four H atoms in the ground electronic state to form a CH_4 molecule. To differentiate this imaginary excitation from the real electronic excitation the former is given the name 'promotion'. The energy necessary for this 'excitation' is more than compensated by the bond energy released upon the formation of the four C-H bonds. The sum of the bond energies amounts to 25.31 eV. This accounts for the formation of the stable CH_4 molecule.

The stabilization accompanying bond formation, in general, will be discussed in detail for the cases of simpler systems such as those of H_2^+ and H_2, in Chaps. 6 and 7. At the moment, it suffices to say that a large energy stabilization is recovered mainly by the exchange interaction between the electrons of the valence state C atom and the four electrons on each H atom. During the approach of the electrons on the H atoms to the electrons on the C atom the two electron spins per each C-H bond are kept anti-parallel (cf. Chaps. 6 and 7).

Pauling (L.C. Pauling, 1931) proposed the concept of sp^3 hybridized orbitals to explain the tetrahedrally directed C-H bonds in the CH_4 molecule. These are virtual orbitals expressed as linear combinations of the atomic orbitals (χ) 2s, $2p_{-1}$, $2p_0$, $2p_{+1}$ (or, alternatively, 2s, $2p_x$, $2p_y$, $2p_z$, ... cf. Sect. 3.3.2). Expressions for the orthonormal sp^3 hybridized orbitals (ϕ) are as follows.

$$\phi_1 = \frac{1}{2}\left(\chi_{2s} + \chi_{2p_x} + \chi_{2p_y} + \chi_{2p_z}\right) \tag{4.9}$$

$$\phi_2 = \frac{1}{2}\left(\chi_{2s} - \chi_{2p_x} - \chi_{2p_y} + \chi_{2p_z}\right) \tag{4.10}$$

$$\phi_3 = \frac{1}{2}\left(\chi_{2s} + \chi_{2p_x} - \chi_{2p_y} - \chi_{2p_z}\right) \tag{4.11}$$

$$\phi_4 = \frac{1}{2}\left(\chi_{2s} - \chi_{2p_x} + \chi_{2p_y} - \chi_{2p_z}\right) \tag{4.12}$$

The fact that ethylene is a planar molecule with four equivalent C-H bonds and that acetylene is linear is also explicable by considering the sp^2 and the sp hybridized orbitals, respectively.

In ethylene, out of the total six sp^2 hybridized orbitals, two are used for the formation of the σ bond between the two C atoms and the remaining four are for the four C-H bonds. As for the two 2p orbitals perpendicular to the molecular plane, they are used to form a π bond. One set of expressions for the orthonormalized sp^2 hybridized orbitals is given below.

$$\phi_1 = \frac{1}{\sqrt{3}}\chi_{2s} + \sqrt{\frac{2}{3}}\chi_{2p_x} \tag{4.13}$$

$$\phi_2 = \frac{1}{\sqrt{3}}\chi_{2s} - \frac{1}{\sqrt{6}}\chi_{2p_x} + \frac{1}{\sqrt{2}}\chi_{2p_y} \tag{4.14}$$

$$\phi_3 = \frac{1}{\sqrt{3}}\chi_{2s} - \frac{1}{\sqrt{6}}\chi_{2p_x} - \frac{1}{\sqrt{2}}\chi_{2p_y} \tag{4.15}$$

In acetylene, one of the 2p orbitals parallel to the molecular axis forms two sp hybridized orbitals by a linear combination with the 2s orbital on each C atom. Out of the total four sp-orbitals in the molecule, two are used for the formation of the σ bond between the two C atoms and the other two are for the formation of the two C-H bonds. The two 2p orbitals on each C atom perpendicular to the molecular axis give the two π bonds. An expression for the two orthonormalized sp hybridized orbitals is as follows.

$$\phi_1 = \frac{1}{\sqrt{2}}\chi_{2s} + \frac{1}{\sqrt{2}}\chi_{2p_x} \tag{4.16}$$

$$\phi_2 = \frac{1}{\sqrt{2}}\chi_{2s} - \frac{1}{\sqrt{2}}\chi_{2p_x} \tag{4.17}$$

Figure 4.5 demonstrates that the minimum excitation energy of $2s^22p^3 \rightarrow 2s2p^4$ for the N atom and $2s^22p^4 \rightarrow 2s2p^5$ for the O atom are about 11 eV and 16 eV, respectively. These energies are significantly larger than the energy for excitation (about 4.2 eV) of $^3P_0 \rightarrow {}^5S^\circ$ for the C atom. This difference in energy explains the difference in the ease of hybridization among the C, N, and O atoms: the lower the excitation energy is, the easier the realization of hybridization is. The fact that the mode of chemical bonding is not so varied in N and O atom-containing molecules relative to C atom-containing molecules stems from the greater difficulty of promotion in the N and O atoms.

Pauling's hybridization concept is general, and not restricted to the three atoms mentioned above. Let us examine the possibility of bond formation involving rare gas atoms. If the He and Ne atoms in the ground electronic configuration are excited as $1s^2 \rightarrow 1s2p$ and $1s^22s^22p^6 \rightarrow 1s^22s^22p^53s$, respectively, they may be expected to behave as divalent atoms and may form some chemical bonds. However, in these cases the promotion energies amount to about 21 eV and 17 eV, respectively, which would hardly be compensated for by the bond energy. Moreover, the increase of the principal quantum numbers of the outermost electron from 1 to 2 for He and from 2 to 3 for Ne results in a considerable diffuseness of the electronic distribution, which would disfavor bond formation. The fact that no stable molecule involving He or Ne atoms is known so far can be attributed to the above reasoning. However, this does not mean that all the rare gas atoms are incapable of yielding molecules. In fact, from the Ar atom on, the rare gas atoms start to produce a variety of molecules under certain conditions. The reason why I described the rare gas atom as *almost inert* in Sect. 4.2 is based on the situation stated here (cf. Sect. 8.4.2 also).

4.4 An Example of the Solution of Multi-Electron Atoms by the Hartree-Fock Method: For the Case of Ne

In Sect. 4.2 it was assumed that each individual electron, e.g., electron 1, in multi-electron atoms obeys the Schrödinger equation (4.6). Without the knowledge of the concrete form of the 'effective' one-electron potential $V(1)$, however, it is impossible to analyze the multi-electron atom quantitatively. In this section we will consider how to solve this problem. As a first step in analyzing the electrostatic interaction between electron 1 and all the other electrons, an approximation technique called the Hartree-Fock method will be introduced. This method is applicable not only to the one-centered multi-electron atom but also to the multi-centered polyatomic molecule to be dealt with in Chaps. 7 through 8.

Since the Hartree-Fock method, in its essence, is the same whether it is applied to multi-electron atoms or to more general systems of polyatomic molecules, details of the method will be described in Chap. 8 (Sects. 8.1 and 8.2) to avoid redundancy. In this section an outline of the method is presented in advance of the later chapter. First, let us have a glance at the Schrödinger equation for polyatomic molecules, which includes the equation for multi-electron atoms as a special case. The equation can be described concisely as follows:

$$\hat{H}\Phi(\boldsymbol{r},\boldsymbol{R}) = E\Phi(\boldsymbol{r},\boldsymbol{R}) \tag{4.18}$$

$$\hat{H} = \hat{T}_{\mathrm{e}} + \hat{T}_{\mathrm{N}} + \hat{V}_{\mathrm{ee}} + \hat{V}_{\mathrm{eN}} + \hat{V}_{\mathrm{NN}} \tag{4.19}$$

Here, the symbols $\boldsymbol{r}$ and $\boldsymbol{R}$ represent inclusively the spatial coordinates of electrons and nuclei. The terms on the right-hand side of (4.19) corresponds to the kinetic energy of electrons, the kinetic energy of nuclei, the interelectronic repulsive energy, the electron-nuclear attractive energy, and the repulsive energy among the nuclei, respectively.

In the problem of multi-electron atoms the term $\hat{V}_{\mathrm{NN}}$ is missing. Our concern in this section is the internal energy of the atoms, which means the energy excluding the kinetic energy of nuclei. Therefore, the Schrödinger equation for multi-electron atoms is written as follows:

$$\hat{H}\Phi(\boldsymbol{r}) = E\Phi(\boldsymbol{r}) \tag{4.20}$$

$$\hat{H} = \hat{T}_{\mathrm{e}} + \hat{V}_{\mathrm{ee}} + \hat{V}_{\mathrm{eN}} \tag{4.21}$$

The eigenfunction $\Phi(\boldsymbol{r})$ for an atom with $N = 2n$ electrons is given in terms of one-electron functions like $\phi_{k_1}(1)$ as follows within the framework of the one-electron approximation described in Sect. 4.1

$$\Phi(1, 2, \ldots, 2n) = \|\phi_{k_1}(1)\phi_{k_2}(2)\ldots\phi_{k_N}(2n)\| \tag{4.5}$$

Our aim here is to obtain these one-electron functions by considering the interelectronic interaction explicitly. The practical procedure for this purpose is the same as the procedure for polyatomic molecules.

When electron 1 is in the i-th one-electron function $\phi_i(1)$, it is determined by the following eigenvalue equation called the Hatree-Fock equation.

$$\hat{F}(1)\phi_i(1) \equiv \left[\hat{H}^0(1) + \sum_{j=1}^{n}(2\hat{J}_j(1) - \hat{K}_j(1)\right]\phi_i(1) = \varepsilon_i\phi_i(1) \tag{4.22}$$

This is the same as (8.9) in Sect. 8.1, where the derivation will be outlined. Here, only the physical significance of the operators in the square brackets in (4.22) will be given.

The first term in the brackets, $\hat{H}^0(1)$, represents the sum of the kinetic energy $-\hbar^2\Delta_1/(2m)$ of electron 1 and the potential energy of this electron in the field of the bare nucleus $-Ze^2/(4\pi\varepsilon_0 r_1)$. The second term consisting of the summation over $j = 1, \ldots, n$ represents the interelectronic interaction energy whose definition will be given in Sect. 8.1.1. It suffices to say here that the operator $\hat{J}_j(1)$ corresponds to the Coulombic repulsion between electron 1 and all the other electrons whereas the operator $\hat{K}_j(1)$ represents the effect of mitigation of the above repulsion when electron 1 and the counterpart electron are in the same spin state. The simple summation over $j = 1, \ldots, n$ implies that electron 1 is assumed to be in an effective potential which has been averaged over all the other electrons. Therefore, this expression in terms of the simple summation may be called the averaged potential field approximation.

As an example of multi-electron atoms, the Ne atom will be examined. As will be described in Sect. 8.1.1, it is not at all an easy task to solve (4.22) even for a relatively simple atom like Ne. Part of the result obtained by C.H. Fischer (1977) is shown in Table 4.3 and Fig. 4.6(a), where the energy eigenvalues and the radial part $R_{n,l}(r)$ of the corresponding three eigenfunctions $\phi_i(1)$ are given.

Since the Hartree-Fock equation is based on the approximation of the averaged potential field, the radial component $R_{n,l}(r)$ is taken to be independent of the azimuthal quantum number l. As a result, it is assumed that the angular component of the one-electron function can be expressed by the familiar spherical surface harmonics. It is noted that in Fig. 4.6(a) the product of r and $R_{n,l}(r)$, rather than $R_{n,l}(r)$ itself, which appeared in (3.18) and

Table 4.3. Orbital energy, ionization energy, and average distance of the electron from the Ne nucleus in each orbital

Orbital	Orbital Energy (ε_i/eV)	Ionization Energy (E_i/eV)	Average Distance ($\langle r \rangle$/pm)
1s	−891.784	860	8.34
2s	−52.529	48.4	47.20
2p	−23.140	21.564	51.08

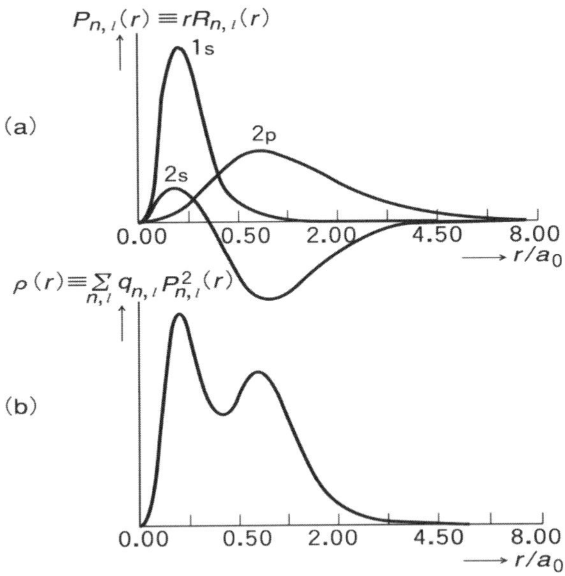

Fig. 4.6. (**a**) Radial component $P_{n,l}(r) \equiv rR_{n,l}(r)$. (**b**) Electron density distribution $\rho(r)$ of the Ne atom

in Table 3.1, is shown. This is because the square of $P_{n,l}(r) \equiv rR_{n,l}$ is proportional to the probability of finding electron 1 when it occupies the orbital $\phi_i(1)$ and is in the space between the two spherical shells of radius r_1 and $r_1 + \mathrm{d}r_1$. On account of the assumption of spherical symmetry the following relation is immediately derived,

$$\int \phi_i^2(1)\mathrm{d}v_1 = 4\pi \int \phi_i^2(1) r_1^2 \mathrm{d}r_1 = 4\pi \int r_1^2 R_{n,l}^2(r)\mathrm{d}r_1 \, .$$

It is known that the energy eigenvalue ε_i is related to the ionization energy E_{i}, which is the minimum energy necessary to remove an electron from its orbital, as follows.

$$\varepsilon_i \cong -E_{\mathrm{i}} \qquad (4.23)$$

This relation is called the Koopmans theorem (some authors prefer the term Koopmans approximation). At any rate, since the ionization energy is physically observable, comparison of the experimental value of the ionization energy with the computed eigenvalue ε_i allows the testing of the Hartree-Fock approximation. As a matter of fact, the fair agreement of $|\varepsilon_i|$ and $|E_\mathrm{i}|$ listed in Table 4.3 indicates that the Hartree-Fock method does work. The limitation of the method will be commented on in Chap. 8.

The average distance in Table 4.3 is calculated by the relation $\langle r \rangle = \int \phi^* r \phi \mathrm{d}v = 4\pi \int r P^2(r) \mathrm{d}r$. It is seen that the electron in the innermost 1s orbital is concentrated near the nucleus and that the average distances of the electrons occupying the 2s and 2p orbitals are similar and are smaller than the Bohr radius. Note that the abscissa of Fig. 4.6 is proportional to the square root of the distance.

Figure 4.6(b) illustrates the radial electron density distribution obtained by summing all the electrons at a certain distance. The symbol $q_{n,l}$ represents the number of electrons occupying the orbital specified by the quantum numbers n and l. Out of the two peaks the inner one comprises mainly the contribution from the 1s orbital and the outer one is mainly due to the 2s and 2p orbitals. The result shown in the figure confirms the concept of the electronic shell structure mentioned in Sect. 4.2. The considerable breadth of the two peaks, however, shows that the illustration of the shells by concentric rings seen in some elementary textbooks is quite misleading.

5 Born-Oppenheimer Approximation: Separation of Electronic Motion from Nuclear Motion in Chemical Bonding

In Chaps. 3 and 4, atomic systems with a single nucleus have been treated. In these systems the nucleus is regarded as motionless because its mass is three to four orders of magnitude larger than the mass of the electron and because we are not concerned with the translational motion of the nucleus. In molecules, however, the motion of both the electrons and the nuclei becomes the subject of our concern, which means that the kinetic energy of the nuclei and the interactions between the nuclei must also be taken into consideration. Intrinsically, the interactions between electrons and nuclei in molecules are inseparable. However, the analysis of the system of electrons and nuclei as a whole is technically a very formidable task. In order to circumvent this difficulty a very useful approximation has been proposed, and it is this approximation technique which serves as the subject of this chapter.

Since the masses of nuclei are several thousand times that of the electron, nuclei move much more slowly than the electron, which allows us to deal with the motions of the two separately. This is what is called the Born-Oppenheimer approximation. By virtue of this approximation, the motion of nuclei can be regarded as being dictated by the potential energy surface, which is determined through the solution of the eigenvalue equation for the electronic energy. In this way the electronic and nuclear states are determined independently.

In general, in terms of energy states, a molecule obeys the following equations, ignoring the electron and nuclear spins:

$$\hat{H}\Phi(\boldsymbol{r}, \boldsymbol{R}) = E\Phi(\boldsymbol{r}, \boldsymbol{R}) \tag{5.1}$$

$$\hat{H} = \hat{T}_{\mathrm{e}} + \hat{T}_{\mathrm{N}} + \hat{V}_{\mathrm{ee}} + \hat{V}_{\mathrm{eN}} + \hat{V}_{\mathrm{NN}} \tag{5.2}$$

These are the same as (4.18) and (4.19), respectively. The first two terms of (5.2) stand for the kinetic energy operators of the electrons (i, j) and nuclei (P, Q), respectively. The following three terms are the electrostatic potential energy operators for the pairs of electron-electron, electron-nucleus, and nucleus-nucleus. They are given explicitly in (5.3) where Z_P and Z_Q stand for the atomic numbers of the nuclei P and Q, and the summations, such as $\sum_{i>j}$ and $\sum_{P>Q}$, are for the different pairs of electrons and nuclei, respectively.

$$\hat{T}_{\mathrm{e}} = \sum_{i} -\hbar^2 \Delta_i/(2m_{\mathrm{e}}), \quad \hat{T}_{\mathrm{N}} = \sum_{p} -\hbar^2 \Delta_P/(2M_P),$$
$$\hat{V}_{\mathrm{ee}} = \sum_{i>j} e^2/(4\pi\varepsilon_0 r_{ij}), \quad \hat{V}_{\mathrm{eN}} = -\sum_{i}\sum_{P} Z_P e^2/(4\pi\varepsilon_0 r_{iP}),$$
$$\hat{V}_{\mathrm{NN}} = \sum_{P>Q} Z_P Z_Q e^2/(4\pi\varepsilon_0 R_{PQ}) \qquad (5.3)$$

So far nothing is known about the energy eigenfunctions of the system $\Phi(\boldsymbol{r}, \boldsymbol{R})$. However, as a guess, let us assume that $\Phi(\boldsymbol{r}, \boldsymbol{R})$ can be decomposed into two parts, one purely dependent on the nuclear coordinates $\Phi_{\mathrm{N}}(\boldsymbol{R})$ and the other on both the electronic and nuclear coordinates $\Phi_{\mathrm{e}}(\boldsymbol{r}, \boldsymbol{R})$. Thus, we have the following expression.

$$\Phi(\boldsymbol{r}, \boldsymbol{R}) \cong \Phi_{\mathrm{N}}(\boldsymbol{R}) \Phi_{\mathrm{e}}(\boldsymbol{r}, \boldsymbol{R}) \qquad (5.4)$$

We are to substitute (5.4) into (5.1), where the Hamiltonian $\hat{H}$ is given by (5.2) and (5.3). The Laplacian in (5.3), Δ, can be rewritten as $\Delta = \nabla\nabla = \nabla^2$ with the symbol ∇ called nabla representing the first order differential (cf. Sect. 1.4.2). Therefore, when the second order differential $\Delta_P = \nabla_P^2$ with regard to the coordinates of nucleus P operates on $\Phi(\boldsymbol{r}, \boldsymbol{R})$, (5.1) can be expanded as follows.

$$\begin{aligned}
&\Phi_{\mathrm{N}}(\boldsymbol{R}) \left[\sum_{i} -\hbar^2 \Delta_i/(2m_{\mathrm{e}}) + \hat{V}_{\mathrm{ee}}(\boldsymbol{r}) + \hat{V}_{\mathrm{eN}}(\boldsymbol{r}, \boldsymbol{R})\right] \Phi_{\mathrm{e}}(\boldsymbol{r}, \boldsymbol{R}) \\
&+\Phi_{\mathrm{e}}(\boldsymbol{r}, \boldsymbol{R}) \left[\sum_{P} -\hbar^2 \Delta_P/(2M_{\mathrm{P}}) + \hat{V}_{\mathrm{NN}}(\boldsymbol{R})\right] \Phi_{\mathrm{N}}(\boldsymbol{R}) \\
&+ \left(\sum_{P} -\hbar^2 \Delta_P/(2M_P)\right) \Phi_{\mathrm{N}}(\boldsymbol{R}) \Delta_P \Phi_{\mathrm{e}}(\boldsymbol{r}, \boldsymbol{R}) \\
&+2 \left(\sum_{P} -\hbar^2 \Delta_P/(2M_P)\right) \nabla_P \Phi_{\mathrm{N}}(\boldsymbol{R}) \nabla_P \Phi_{\mathrm{e}}(\boldsymbol{r}, \boldsymbol{R}) \\
&\cong E \Phi_{\mathrm{N}}(\boldsymbol{R}) \Phi_{\mathrm{e}}(\boldsymbol{r}, \boldsymbol{R})
\end{aligned} \qquad (5.5)$$

What is sought are the two factors $\Phi_{\mathrm{N}}(\boldsymbol{R})$ and $\Phi_{\mathrm{e}}(\boldsymbol{r}, \boldsymbol{R})$ in (5.4). Obviously, however, the single equation (5.5) can not give the two unknown factors independently. This problem, however, was overcome by M. Born and J.R. Oppenheimer (1927), who noticed that the technique of separation of variables introduced in Chaps. 2 and 3 can be applied to obtain both $\Phi_{\mathrm{N}}(\boldsymbol{R})$ and $\Phi_{\mathrm{e}}(\boldsymbol{r}, \boldsymbol{R})$ independently, in so far as the third and fourth terms on the left-hand side of (5.5) are negligibly small compared with the first and second terms. When the former two terms are negligible, the remaining two terms lead to the following relation after division of both sides by the product $\Phi_{\mathrm{N}}(\boldsymbol{R}) \Phi_{\mathrm{e}}(\boldsymbol{r}, \boldsymbol{R})$.

$$\left[\sum_i -\hbar^2 \Delta_i/(2m_\mathrm{e}) + \hat{V}_\mathrm{ee}(\boldsymbol{r}) + \hat{V}_\mathrm{eN}(\boldsymbol{r},\boldsymbol{R})\right] \Phi_\mathrm{e}(\boldsymbol{r},\boldsymbol{R})/\Phi_\mathrm{e}(\boldsymbol{r},\boldsymbol{R})$$
$$+\left[\sum_P -\hbar^2 \Delta_P/(2M_P) + \hat{V}_\mathrm{NN}(\boldsymbol{R})\right] \Phi_\mathrm{N}(\boldsymbol{R})/\Phi_\mathrm{N}(\boldsymbol{R}) \cong E \tag{5.6}$$

The second term on the left-hand side involves only the nuclear coordinates while the first term depends on both the nuclear and electronic coordinates. In order for the sum of these terms to be equal identically to the constant on the right-hand side, the two terms must be individually constant. When the constant for the first term is written as $E_\mathrm{e}(\boldsymbol{R})$ to emphasize that the constant depends on the nuclear coordinate parametrically, the constant for the second term must be equated to $E - E_\mathrm{e}(\boldsymbol{R})$.

The three terms in the first square brackets on the left-hand side of (5.6) altogether represent the electronic energy operator. Denoting this operator by $\hat{H}_\mathrm{e}$, the following eigenvalue equation is obtained for the electronic part.

$$\hat{H}_\mathrm{e}\Phi_\mathrm{e}(\boldsymbol{r},\boldsymbol{R}) = \left(\hat{T}_\mathrm{e} + \hat{V}_\mathrm{ee} + \hat{V}_\mathrm{eN}\right)\Phi_\mathrm{e}(\boldsymbol{r},\boldsymbol{R}) = E_\mathrm{e}(\boldsymbol{R})\Phi_\mathrm{e}(\boldsymbol{r},\boldsymbol{R}) \tag{5.7}$$

Equation (5.7) shows that the constant $E_\mathrm{e}(\boldsymbol{R})$ introduced as a constant in the method of separation of variables has a physical meaning as the eigenvalue of the eigenvalue equation for the electronic energy. Similarly, by denoting the second square brackets on the left-hand side of (5.6) by $\hat{H}_\mathrm{N}$, another eigenvalue equation for the nuclear part is obtained as follows after a slight rearrangement.

$$\left[\hat{H}_\mathrm{N} + E_\mathrm{e}(\boldsymbol{R})\right]\Phi_\mathrm{N}(\boldsymbol{R}) = \left[\hat{T}_\mathrm{N} + \hat{V}_\mathrm{NN} + E_\mathrm{e}(\boldsymbol{R})\right]\Phi_\mathrm{N}(\boldsymbol{R}) \cong E\Phi_\mathrm{N}(\boldsymbol{R}) \tag{5.8}$$

Equation (5.8) is interpreted as an approximate eigenvalue equation for the total energy E of a molecule in which the nuclei are moving in an effective potential field $\hat{V}_\mathrm{NN} + E_\mathrm{e}(\boldsymbol{R})$.

By solving both (5.7) and (5.8), the eigenvalue and eigenfunction of the original eigenvalue equation of (5.1) can be obtained approximately. This is the Born-Oppenheimer approximation. In the study of chemical bonding the electronic eigenfunction $\Phi_\mathrm{e}(\boldsymbol{r},\boldsymbol{R})$ appearing in (5.4) through (5.7) is assumed to be the function that describes the *ground electronic state* of the molecule under investigation. In other words, any possible effect of excited electronic states on chemical bonding is usually ignored. For this reason the Born-Oppenheimer approximation is sometimes called the adiabatic approximation*. However, this somewhat abstract terminology must be used with care because (5.4) is an approximate equation and the extent of the approximations may differ from one author to the other. To be more specific, the total eigenfunction in (5.4) could have been rigorously related to the electronic eigenfunctions as in (5.9) below because the electronic eigenfunctions satisfy the eigenvalue equation of (5.7) so that the completeness of the electronic eigenfunctions allows the exact expression of (5.9) (cf. Sect. 2.1.10).

$$\Phi(\boldsymbol{r}, \boldsymbol{R}) = \sum_{i=1}^{\infty} \Phi_{\mathrm{N}}(\boldsymbol{R}) \Phi_{\mathrm{e}i}(\boldsymbol{r}, \boldsymbol{R}) \tag{5.9}$$

In (5.4), however, only the ground electronic state, $\Phi_{\mathrm{e}i=1}(\boldsymbol{r}, \boldsymbol{R})$, is taken into account so that I have used the symbol $\cong$ in that equation. Depending upon how many electronic states are included in the sum of (5.9), the meaning of 'adiabatic' as mentioned above changes. A related matter as concerns this adiabaticity will be discussed briefly in Sect. 8.3.1.

We now proceed to the procedure for utilizing the Born-Oppenheimer approximation in solving a practical problem involving molecules. First, the nuclear coordinates $\boldsymbol{R}$ in (5.7) are fixed at a certain value to obtain (5.10). Then, the eigenvalue $E_{\mathrm{e}}(\boldsymbol{R}_{\mathrm{fixed}})$ and the corresponding eigenfunction $\Phi_{\mathrm{e}}(\boldsymbol{r}, \boldsymbol{R}_{\mathrm{fixed}})$ are solved for.

$$\hat{H}_{\mathrm{e}} \Phi_{\mathrm{e}}(\boldsymbol{r}, \boldsymbol{R}_{\mathrm{fixed}}) = E_{\mathrm{e}}(\boldsymbol{R}_{\mathrm{fixed}}) \Phi_{\mathrm{e}}(\boldsymbol{r}, \boldsymbol{R}_{\mathrm{fixed}}) \tag{5.10}$$

The solution $E_{\mathrm{e}}(\boldsymbol{R}_{\mathrm{fixed}})$ is, in turn, substituted to (5.8) to obtain the approximate eigenvalue E and the corresponding eigenfunction $\Phi_{\mathrm{N}}(\boldsymbol{R})$. Since the nuclear coordinate is fixed, the term $\hat{V}_{\mathrm{NN}}$ in (5.8) is merely a constant equal to $V_{\mathrm{NN}} = \sum_{P>Q} Z_P Z_Q e^2/(4\pi\varepsilon_0 R_{PQ,\mathrm{fixed}})$.

In diatomic molecules, the nuclear coordinate is solely the internuclear distance R. By the procedure described above the energy $E_{\mathrm{e}}(R_{\mathrm{fixed}})$ is determined at various values of R_{fixed}, and by adding the nuclear repulsion term $V_{\mathrm{NN}} = Z_P Z_Q e^2/(4\pi\varepsilon_0 R_{PQ,\mathrm{fixed}})$ to $E_{\mathrm{e}}(R_{\mathrm{fixed}})$, the effective potential energy operator for the nuclear motion is obtained as $\hat{V}_{\mathrm{eff}}(R) = E_{\mathrm{e}}(R) + V_{\mathrm{NN}}(R)$.

Figure 5.1 demonstrates schematically the potential energy function $V_{\mathrm{eff}}(R) = E_{\mathrm{e}}(R) + V_{\mathrm{NN}}(R)$. The two components on the right-hand side are individually shown by the vertical arrows at two internuclear distances R_1 and R_2.

Since energy is a relative quantity in chemistry, the origin of the ordinate in Fig. 5.1 can be chosen arbitrarily. In the figure the origin of the total energy E of the system is taken when the two atoms with the nuclear charges of $Z_P e$ and $Z_Q e$ are separated infinitely. The internuclear distance corresponding to the minimum of the effective potential curve $V_{\mathrm{eff}}(R)$ is called the equilibrium internuclear distance. The bold horizontal line just above the minimum of the curve represents the first allowed energy level for the nuclei, which represents the zero-point energy level (as for the zero-point energy for the electron, cf. Sect. 1.2).

From the zero-point energy level up to the level of $E = 0$ there are shown schematically a number of horizontal lines, which represent the energy levels of quantized states for the nuclear rotation and vibration. These energy levels of various values of E are the subject of ro-vibrational spectroscopy of molecules.

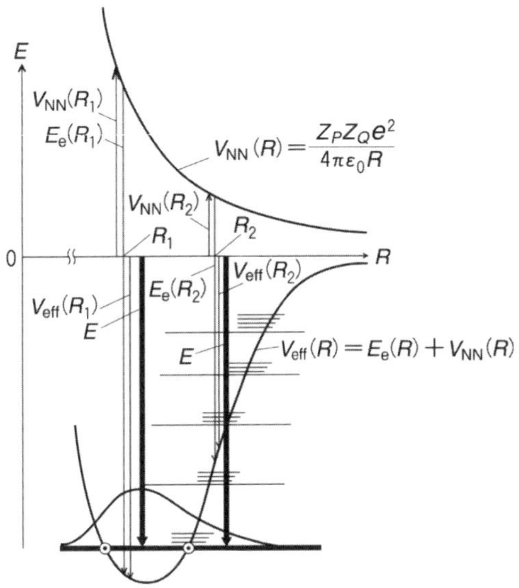

Fig. 5.1. Schematic potential energy function for diatomic molecules

The bell-shaped curve is a schematic illustration of the eigenfunction for the zero-point energy level of the nuclei. The two points at the intersections of the horizontal line and the potential curve shown by the open circles indicate the approximate confinement range of the nuclei. The extension of the eigenfunction outside these circles indicates that the nuclear motion is allowed even beyond the classically forbidden region. This extension of the eigenfunction is characteristic of quantum mechanics, and is called the tunnel effect. Similar behavior occurs in all the ro-vibrationally excited energy levels demonstrated by the thin horizontal lines.

Figure 5.1 is valid for the molecular ground electronic state case. Similar diagrams are obtained for each excited electronic state, including the ionized states. While the diagram for the ground electronic state is essential in the study of chemical bonding, those for the excited electronic states are the main subject of electronic spectroscopy and photochemistry.

Next, we will examine the accuracy of the Born-Oppenheimer approximation in light of the omission of the third and fourth terms in (5.5) in the derivation of (5.6). The omission of thee terms is tantamount to the assumption that $\nabla_P \Phi_N(\boldsymbol{R}) \gg \nabla_P \Phi_e(\boldsymbol{r}, \boldsymbol{R})$; the factor $\nabla_P \Phi_e(\boldsymbol{r}, \boldsymbol{R})$ in the fourth term of (5.5) represents the rate of change of $\Phi_e(\boldsymbol{r}, \boldsymbol{R})$ with respect to the nuclear motion, which is much slower than the electronic motion. This means that the light electrons can quickly catch up with the nuclear motion. As a result, the increment of $\Phi_e(\boldsymbol{r}, \boldsymbol{R})$ during the change from $\boldsymbol{R}$ to $\boldsymbol{R}+\delta\boldsymbol{R}$ can be considered to be sufficiently small. As for the factor $\Delta_P \Phi_e(\boldsymbol{r}, \boldsymbol{R}) \equiv \nabla_P \nabla_P \Phi_e(\boldsymbol{r}, \boldsymbol{R})$ in the third term of (5.5), it is even smaller because of the dual difference in $\Phi_e(\boldsymbol{r}, \boldsymbol{R})$.

Another explanation in support of the adequacy of the approximation is as follows. In ordinary molecular vibrations the motions of the nuclei are confined to small regions near their equilibrium positions (roughly speaking the vibrational amplitude of nuclei is only about 1% of typical equilibrium internuclear distances). In contrast to this restriction on the nuclear motion, the electrons in molecules are allowed to move over the whole region of the molecule. Under such a circumstance the effect of a small displacement of the nuclei on $\Phi_{\mathrm{N}}(\boldsymbol{R})$, i.e., the differential $\nabla_P \Phi_{\mathrm{N}}(\boldsymbol{R})$ should be much larger than the effect on $\Phi_{\mathrm{e}}(\boldsymbol{r}, \boldsymbol{R})$, which is measured by $\nabla_P \Phi_{\mathrm{e}}(\boldsymbol{r}, \boldsymbol{R})$. The above assumption of the inequality, $\nabla_P \Phi_{\mathrm{N}}(\boldsymbol{R}) \gg \nabla_P \Phi_{\mathrm{e}}(\boldsymbol{r}, \boldsymbol{R})$, is interpreted as expressing this difference in the effect of the nuclear displacement.

It should be remembered, however, that the Born-Oppenheimer approximation fails fatally when the electronic state determined by the eigenvalue equation (5.7) is degenerate (or nearly degenerate). In such a case the nuclear motion is sensitively affected by the electronic motion and the equilibrium nuclear arrangement may be conspicuously deformed. The effect of causing a distortion from a highly symmetric nonlinear equilibrium arrangement is called the Jahn-Teller effect*. A somewhat similar Renner-Teller effect* occurs in linear molecules.

With the exceptions of the cases just mentioned, the analysis of most problems concerning molecules are treated on the basis of the Born-Oppenheiner approximation. Without this assumption, the exact solution would be beyond our reach except for a very limited number of simple systems.

Problem

5.1 Verify (5.5).

6 The Hydrogen Molecular Ion: The Simplest, but the Most Fundamental System for Understanding Chemical Bonds

The simplest molecule among all is the hydrogen molecular ion H_2^+. The ion can be obtained by ionizing H_2 molecules through electron bombardment, for example. In the presence of neutral H_2 molecules, the ion is subjected to a fast chemical reaction, $H_2^+ + H_2 \rightarrow H_3^+ + H$. If this reaction is suspended by some means, the thermodynamically stable H_2^+ ion can be observed spectroscopically. Through such observation, the ion is found to be stable by as much as $269\,\mathrm{kJ\,mol^{-1}} = 2.79\,\mathrm{eV}$, relative to the dissociation limit of $H + H^+$.

In the H_2^+ ion, two repulsive protons share one electron with remarkable stability. The purpose of this chapter is to clarify the mechanism of this stability. In the end, it will be shown that the main reason for the stability is that the two states $H^+ - H$ and $H - H^+$, are intrinsically indistinguishable irrespective of the distance of the two protons. Such a situation is described as an exchange degeneracy.

The concept of the degeneracy of a state is very proper to quantum mechanics. This concept has to do with the view of W. Heitler and F. London (1927), who attributed the essential cause of the bonding in the H_2 molecule to a 'quantum mechanical beat phenomenon' (cf. Chap. 7).

This quantum mechanical phenomenon advocated by Heitler and London was associated later with what is called the exchange force. However, as G. Herzberg (1989), among others, pointed out in his noted book on diatomic molecules, the exchange force in atoms and molecules is not a new mysterious force but is a force originating from the familiar electrostatic interaction between electrons and nuclei. What is important is that in quantum mechanics the superposition of states, which are indistinguishable in classical mechanics, has a fundamental importance. In this chapter the essential features of bonding in the H_2^+ ion will be explained in several ways.

6.1 The Stability of the H_2^+ Ion (Explanation 1): Explanation by Invoking the Wave Packet of the Electron

First, we will consider the symmetry of the electronic eigenfunction in the ion, and then construct the electron wave packet for the ion. The stability of bonding in the ion is, then, attributed to the traveling of the wave packet back and forth between the two protons.

6.2 The Stability of the H_2^+ Ion (Explanation 2): Qualitative Explanation in Terms of the Uncertainty Principle
A qualitative explanation by Feynman will be introduced briefly.

6.3 The Stability of the H_2^+ Ion (Explanation 2): An Explanation in Terms of the LCAO-MO Method
The H_2^+ ion is the starting point for the LCAO-MO method. Two approaches are introduced. One is to start with a consideration of the symmetry of the molecular orbital, the other is based on Ritz's method of calculus of variation.

6.4 A Few Topics in the Study of the H_2^+ System
The simplest molecule still attracts interest in both experiment and theory.

6.1 The Stability of the H_2^+ Ion (Explanation 1): Explanation by Invoking the Wave Packet of the Electron

The fact that the H_2^+ ion in its ground electronic state exists stably in spite of the repulsion between the two protons has attracted the attention not only of chemists but also of some physicists. Nuclear physicists consider the exchange of elementary particles between protons and neutrons in nuclei to explain the huge nuclear energy of 2.225 MeV. There is some analogy between the exchange of elementary particles in the deuteron and the exchange of an electron in the $H + H^+$ system, as pointed out by S. Tomonaga (1974). He stated in his book that "Heisenberg cut to the essence of the nuclear force and thought it must be an *exchange force*. The concept of exchange force appears in the field of chemistry. For example, when two hydrogen atoms combine to form a hydrogen molecule, the force which binds the atoms is an exchange force. ...Now let me talk about the process occurring when two hydrogen atoms attract one another to combine into a hydrogen molecule and why we call the force responsible for it an exchange force. Since the case of H_2^+ ion is simpler than the case of hydrogen molecule and is closer to that of a neutron and a proton, let me use it as an example as we proceed".

After a lucid and simple explanation of the bonding in the H_2^+ ion Tomonaga concluded in his book that "We therefore can say the exchange force originates in the shuttling of the electron; ...." (quoted from Oka's translation (1997) of Tomonaga's original in Japanese (1974)). R.P. Feynman, who shared the Nobel prize in physics in 1965 with Tomonaga, also compared the H_2^+ ion and deuteron for the explanation of the attractive forces in them (1965).

In the following I will introduce the explanation by Tomonaga in terms of 'the shuttling of the electron'. In this introduction, I will first consider a general feature of the eigenfunction of the H_2^+ ion in Sect. 6.1.1 and then proceed to 6.1.2 where 'the shuttling of the electron' is described.

6.1.1 The Eigenfunction of the H_2^+ Ion with the Proper Symmetry

Let us consider that two protons are separated by a distance of R, as shown in Fig. 6.1, and an electron is located at the position determined by the vector $\boldsymbol{r}$. The coordinates of the electron relative to the two protons at the left and right are $\boldsymbol{r} - (-\boldsymbol{R}/2) = \boldsymbol{r} + \boldsymbol{R}/2$ and $\boldsymbol{r} - (+\boldsymbol{R}/2) = \boldsymbol{r} - \boldsymbol{R}/2$, respectively. When R is large enough, the electron is regarded as being almost in the vicinity of either one of the two protons.

Such an electron should be approximately described by the 1s orbital of the H atom, i.e., $\chi_l(\boldsymbol{r} + \boldsymbol{R}/2)$ nearly equal to $1\mathrm{s}(\boldsymbol{r} + \boldsymbol{R}/2)$ and $\chi_r(\boldsymbol{r} - \boldsymbol{R}/2)$ nearly equal to $1\mathrm{s}(\boldsymbol{r} - \boldsymbol{R}/2)$. When R is infinitely large, the electron is associated with either one of the two protons, which is described exactly as $\chi(\boldsymbol{r}) = 1\mathrm{s}(\boldsymbol{r})$, with $\boldsymbol{r}$ representing the electron coordinate from the associated proton.

In the H_2^+ ion, R is finite and the eigenfunction of the system obeys the following eigenvalue equation.

$$\begin{aligned}
&\hat{H}(\boldsymbol{r}, \boldsymbol{R})\Phi(\boldsymbol{r}, \boldsymbol{R}) \\
&= \left[\frac{-\hbar^2}{2m_\mathrm{e}}\Delta - \frac{e^2}{4\pi\varepsilon_0|\boldsymbol{r} + \boldsymbol{R}/2|} - \frac{e^2}{4\pi\varepsilon_0|\boldsymbol{r} - \boldsymbol{R}/2|} + \frac{e^2}{4\pi\varepsilon_0 R}\right]\Phi(\boldsymbol{r}, \boldsymbol{R}) \\
&= E\Phi(\boldsymbol{r}, \boldsymbol{R})
\end{aligned} \tag{6.1}$$

Upon the reversal of the sign of $\boldsymbol{r}$ we have the equation $\hat{H}(-\boldsymbol{r}, \boldsymbol{R})\Phi(-\boldsymbol{r}, \boldsymbol{R}) = E\Phi(-\boldsymbol{r}, \boldsymbol{R})$ from (6.1). Comparison of $\hat{H}(\boldsymbol{r}, \boldsymbol{R})$ and $\hat{H}(-\boldsymbol{r}, \boldsymbol{R})$ shows that the reversal does not change the sum in the square brackets above, which means that both $\Phi(\boldsymbol{r}, \boldsymbol{R})$ and $\Phi(-\boldsymbol{r}, \boldsymbol{R})$ are eigenfunctions of the same Hamiltonian with the same eigenvalue.

When the reversal is repeated twice, the eigenfunction resumes the original form. Thus, by the same logic as that for the exchange of two electrons stated in Sect. 4.1, if the eigenfunction $\Phi(\boldsymbol{r}, \boldsymbol{R})$ is non-degenerate, it must behave either as $\Phi(\boldsymbol{r}, \boldsymbol{R}) = +\Phi(-\boldsymbol{r}, \boldsymbol{R})$ or as $\Phi(\boldsymbol{r}, \boldsymbol{R}) = -\Phi(-\boldsymbol{r}, \boldsymbol{R})$. In other words, with the exchange $\boldsymbol{r} \leftrightarrow -\boldsymbol{r}$ the eigenfunction of the system $\Phi(\boldsymbol{r}, \boldsymbol{R})$

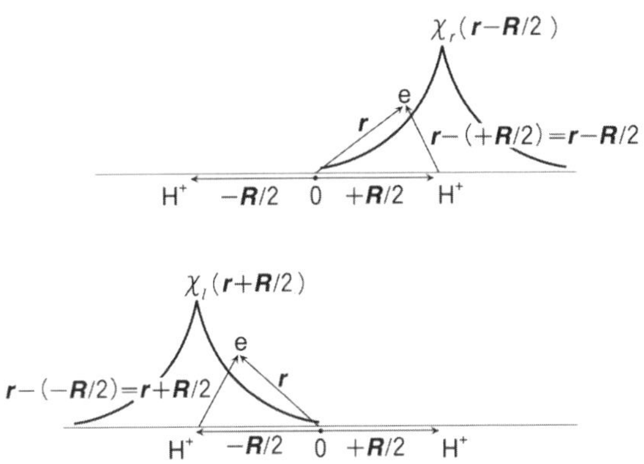

Fig. 6.1. The coordinate system for the H_2^+ ion

must be either symmetric as $\Phi(\boldsymbol{r},\boldsymbol{R}) = +\Phi(-\boldsymbol{r},\boldsymbol{R}) \equiv \Phi_{\rm s}(\boldsymbol{r},\boldsymbol{R})$, or anti-symmetric as $\Phi(\boldsymbol{r},\boldsymbol{R}) = -\Phi(-\boldsymbol{r},\boldsymbol{R}) \equiv \Phi_{\rm a}(\boldsymbol{r},\boldsymbol{R})$. This requirement must hold irrespective of the magnitude of R.

As mentioned above, when R is sufficiently large, the electron is approximated either by $\chi_l(\boldsymbol{r}+\boldsymbol{R}/2) \cong 1{\rm s}(\boldsymbol{r}+\boldsymbol{R}/2)$ or by $\chi_r(\boldsymbol{r}-\boldsymbol{R}/2) \cong 1{\rm s}(\boldsymbol{r}-\boldsymbol{R}/2)$. Both $\chi_l(\boldsymbol{r}+\boldsymbol{R}/2)$ and $\chi_r(\boldsymbol{r}-\boldsymbol{R}/2)$ are degenerate in energy. This degeneracy is named the exchange degeneracy. Under such circumstances any linear combination of $\chi_l(\boldsymbol{r}+\boldsymbol{R}/2)$ and $\chi_r(\boldsymbol{r}-\boldsymbol{R}/2)$, i.e., $p_k\chi_l(\boldsymbol{r}+\boldsymbol{R}/2)+q_k\chi_r(\boldsymbol{r}-\boldsymbol{R}/2)$ with p_k and q_k being the coefficients discussed in Sect. 2.1.14, should be a good approximation of the eigenfunction of the system.

With the above preparation we are ready to search for the symmetry adapted eigenfunctions $\Phi_{\rm s}(\boldsymbol{r},\boldsymbol{R})$ and $\Phi_{\rm a}(\boldsymbol{r},\boldsymbol{R})$ by the perturbation method for degenerate systems described in Sect. 2.1.14.

We first rewrite (6.1) as (6.2) below, which represents the degenerate states with the electron being mainly localized on the left and right protons, but perturbed by the opposite protons.

$$\begin{aligned}
\hat{H} &= \left[\frac{-\hbar^2}{2m_{\rm e}}\Delta - \frac{e^2}{4\pi\varepsilon_0|\boldsymbol{r}+\boldsymbol{R}/2|} + \frac{e^2}{4\pi\varepsilon_0 R}\right] - \frac{e^2}{4\pi\varepsilon_0|\boldsymbol{r}-\boldsymbol{R}/2|} \\
&\equiv \hat{H}_l^0 + \hat{H}_l' \\
\hat{H} &= \left[\frac{-\hbar^2}{2m_{\rm e}}\Delta - \frac{e^2}{4\pi\varepsilon_0|\boldsymbol{r}-\boldsymbol{R}/2|} + \frac{e^2}{4\pi\varepsilon_0 R}\right] - \frac{e^2}{4\pi\varepsilon_0|\boldsymbol{r}+\boldsymbol{R}/2|} \\
&\equiv \hat{H}_r^0 + \hat{H}_r'
\end{aligned} \tag{6.2}$$

We now seek values for p_k and q_k which satisfy the relationship, $\hat{H}\{p_k\chi_l(\boldsymbol{r}+\boldsymbol{R}/2)+q_k\chi_r(\boldsymbol{r}-\boldsymbol{R}/2)\} = E\{p_k\chi_l(\boldsymbol{r}+\boldsymbol{R}/2)+q_k\chi_r(\boldsymbol{r}-\boldsymbol{R}/2)\}$ according to the statement in Sect. 2.1.14. This relationship can be rewritten as in (6.3) below in reference to (6.2).

$$\begin{aligned}
&\hat{H}\{p_k\chi_l(\boldsymbol{r}+\boldsymbol{R}/2)+q_k\chi_r(\boldsymbol{r}-\boldsymbol{R}/2)\} \\
&\equiv p_k(\hat{H}_l^0+\hat{H}_l')\chi_l(\boldsymbol{r}+\boldsymbol{R}/2)+q_k(\hat{H}_r^0+\hat{H}_r')\chi_r(\boldsymbol{r}-\boldsymbol{R}/2) \\
&= p_k(E_{1\rm s}+\hat{H}_l')\chi_l(\boldsymbol{r}+\boldsymbol{R}/2)+q_k(E_{1\rm s}+\hat{H}_r')\chi_r(\boldsymbol{r}-\boldsymbol{R}/2) \\
&= E\{p_k\chi_l(\boldsymbol{r}+\boldsymbol{R}/2)+q_k\chi_r(\boldsymbol{r}-\boldsymbol{R}/2)\}
\end{aligned} \tag{6.3}$$

The constant $E_{1\rm s}$ stands for the energy of an isolated H atom in its ground electronic state, which satisfies the relations of $\hat{H}_l^0\chi_l(\boldsymbol{r}+\boldsymbol{R}/2) = E_{1\rm s}\chi_l(\boldsymbol{r}+\boldsymbol{R}/2)$ and $\hat{H}_r^0\chi_r(\boldsymbol{r}-\boldsymbol{R}/2) = E_{1\rm s}\chi_r(\boldsymbol{r}-\boldsymbol{R}/2)$. Multiplying $\chi_l^*(\boldsymbol{r}+\boldsymbol{R}/2)$ from the left to the last two sides of (6.3) and integrating over the whole space, we have the first of (6.4) below. Similar multiplication of $\chi_r^*(\boldsymbol{r}-\boldsymbol{R}/2)$ followed by integration leads to the second of (6.4):

$$
\begin{aligned}
&p_k \left[\int \chi_l^*(\boldsymbol{r}+\boldsymbol{R}/2) \left\{ \hat{H}_l' - (E - E_{1s}) \right\} \chi_l(\boldsymbol{r}+\boldsymbol{R}/2) \mathrm{d}v \right] \\
&\quad + q_k \left[\int \chi_l^*(\boldsymbol{r}+\boldsymbol{R}/2) \left\{ \hat{H}_r' - (E - E_{1s}) \right\} \chi_r(\boldsymbol{r}-\boldsymbol{R}/2) \mathrm{d}v \right] = 0 \\
&p_k \left[\int \chi_r^*(\boldsymbol{r}-\boldsymbol{R}/2) \left\{ \hat{H}_l' - (E - E_{1s}) \right\} \chi_l(\boldsymbol{r}+\boldsymbol{R}/2) \mathrm{d}v \right] \\
&\quad + q_k \left[\int \chi_r^*(\boldsymbol{r}-\boldsymbol{R}/2) \left\{ \hat{H}_r' - (E - E_{1s}) \right\} \chi_r(\boldsymbol{r}-\boldsymbol{R}/2) \mathrm{d}v \right] = 0
\end{aligned}
\tag{6.4}
$$

Rewriting $E - E_{1s}$ in (6.4) as $E_{k_1}^1$ and introducing the overlap integral $S(R)$ as $S(R) = \int \chi_l^*(\boldsymbol{r}+\boldsymbol{R}/2)\chi_r(\boldsymbol{r}-\boldsymbol{R}/2)\mathrm{d}v = \int \chi_r^*(\boldsymbol{r}-\boldsymbol{R}/2)\chi_l(\boldsymbol{r}+\boldsymbol{R}/2)\mathrm{d}v$ which is put to be equal to 0 because of the assumed large value of R, we have the following secular equation (6.5), which corresponds to the general secular equation (2.54) for the set of (2.52) and (2.53) (cf. Sect. 2.1.14).

$$
\begin{vmatrix}
\int \chi_l^*(\boldsymbol{r}+\boldsymbol{R}/2)\hat{H}_l'\chi_l(\boldsymbol{r}+\boldsymbol{R}/2)\mathrm{d}v - E_k^1 & \int \chi_l^*(\boldsymbol{r}+\boldsymbol{R}/2)\hat{H}_r'\chi_r(\boldsymbol{r}-\boldsymbol{R}/2)\mathrm{d}v \\
\int \chi_r^*(\boldsymbol{r}-\boldsymbol{R}/2)\hat{H}_l'\chi_l(\boldsymbol{r}+\boldsymbol{R}/2)\mathrm{d}v & \int \chi_r^*(\boldsymbol{r}-\boldsymbol{R}/2)\hat{H}_r'\chi_r(\boldsymbol{r}-\boldsymbol{R}/2)\mathrm{d}v - E_k^1
\end{vmatrix}
= 0
\tag{6.5}
$$

The above secular equation has the form, $\begin{vmatrix} \alpha - E & \beta \\ \beta & \alpha - E \end{vmatrix} = 0$, when the symmetry of the system is considered, where $\alpha \equiv \int \chi_l^*(\boldsymbol{r}+\boldsymbol{R}/2)\hat{H}_l'\chi_l(\boldsymbol{r}+\boldsymbol{R}/2)\mathrm{d}v = \int \chi_r^*(\boldsymbol{r}-\boldsymbol{R}/2)\hat{H}_r'\chi_r(\boldsymbol{r}-\boldsymbol{R}/2)\mathrm{d}v$ and $\beta \equiv \int \chi_l^*(\boldsymbol{r}+\boldsymbol{R}/2)\hat{H}_r'\chi_r(\boldsymbol{r}-\boldsymbol{R}/2)\mathrm{d}v = \int \chi_r^*(\boldsymbol{r}-\boldsymbol{R}/2)\hat{H}_l'\chi_l(\boldsymbol{r}+\boldsymbol{R}/2)\mathrm{d}v$.

It is noted that the above approximation of neglecting the overlap integral in (6.4) is similar to the approximation commonly adopted in elementary quantum chemistry treatments by the so-called Simple Hückel Molecular Orbital Theory.

Equation (6.5) is immediately solved to obtain the following two normalized functions to the zero-th order approximation.

$$
\begin{aligned}
\Phi_s(\boldsymbol{r},\boldsymbol{R}) &\equiv \Phi_s = [\chi_l(\boldsymbol{r}+\boldsymbol{R}/2) + \chi_r(\boldsymbol{r}-\boldsymbol{R}/2)] / \sqrt{2(1+S)} \\
\Phi_a(\boldsymbol{r},\boldsymbol{R}) &\equiv \Phi_a = [\chi_l(\boldsymbol{r}+\boldsymbol{R}/2) - \chi_r(\boldsymbol{r}-\boldsymbol{R}/2)] / \sqrt{2(1-S)}
\end{aligned}
\tag{6.6}
$$

The two functions are symmetric and anti-symmetric because a change $\boldsymbol{r} \to -\boldsymbol{r}$ causes the change into $[\chi_l(-\boldsymbol{r}+\boldsymbol{R}/2) + \chi_r(-\boldsymbol{r}-\boldsymbol{R}/2)]$ in Φ_s, which is identical with $[\chi_r(\boldsymbol{r}-\boldsymbol{R}/2) + \chi_l(\boldsymbol{r}+\boldsymbol{R}/2)]$, whereas a change $\boldsymbol{r} \to -\boldsymbol{r}$ in Φ_a results in $[\chi_l(-\boldsymbol{r}+\boldsymbol{R}/2) - \chi_r(-\boldsymbol{r}-\boldsymbol{R}/2)] \to [\chi_r(\boldsymbol{r}-\boldsymbol{R}/2) - \chi_l(\boldsymbol{r}+\boldsymbol{R}/2)]$ which is equal to $-[\chi_l(\boldsymbol{r}+\boldsymbol{R}/2) - \chi_r(\boldsymbol{r}-\boldsymbol{R}/2)]$.

The above result, summarized in (6.6), has been achieved for a nearly degenerate system for large values of R. The two functions in (6.6), however, behave properly with the exchange $\boldsymbol{r} \leftrightarrow -\boldsymbol{r}$ irrespective of the distance R between the two protons. At the same time, the functions are only approximate because the 1s orbital of the H atom must be distorted by the approaching counterpart proton.

Aside from this obvious fault, the importance of the functional form of Φ_s and Φ_a in (6.6) cannot be overemphasized because they describe correctly the equal participation of the two orbitals *and* satisfy the requirement of the symmetry.

Once the eigenfunctions are given, the eigenvalues associated with the functions are obtained by the following integrals.

$$E_\mathrm{s} = \int \Phi_\mathrm{s}^* \hat{H} \Phi_\mathrm{s} \mathrm{d}v, \quad E_\mathrm{a} = \int \Phi_\mathrm{a}^* \hat{H} \Phi_\mathrm{a} \mathrm{d}v \tag{6.7}$$

Fortunately, the above integrals can be performed elementarily by expressing the eigenfunctions $\chi_l(\boldsymbol{r} + \boldsymbol{R}/2)$ and $\chi_r(\boldsymbol{r} - \boldsymbol{R}/2)$ in terms of the 1s orbital of the H atom (cf. Sect. 6.3.3) in the confocal elliptical coordinate (cf. Fig. 6.5).

The results of the integrals are given in (6.23) and (6.27) in Sect. 6.3.3. At the moment, we focus on the dependence of the eigenvalues on large internuclear distances R. From (6.27) to be given in Sect. 6.3.3, the following results are obtained, where $E_{1\mathrm{s}}$ stands for the energy of the 1s orbital of an isolated H atom.

$$\begin{aligned} E_\mathrm{s} &\approx E_{1\mathrm{s}} - \frac{e^2}{4\pi\varepsilon_0 a_0}\left(1 + \frac{R}{a_0}\right)\mathrm{e}^{-R/a_0} \\ E_\mathrm{a} &\approx E_{1\mathrm{s}} + \frac{e^2}{4\pi\varepsilon_0 a_0}\left(1 + \frac{R}{a_0}\right)\mathrm{e}^{-R/a_0} \end{aligned} \tag{6.8}$$

$$S = \left[1 + \frac{R}{a_0} + \frac{R^2}{3a_0^2}\right]\mathrm{e}^{-R/a_0} \tag{6.9}$$

Importantly, the first of (6.8) shows that the energy E_s for the symmetric eigenfunction Φ_s decreases with a decrease in R, which gives rise to an attractive force for the formation of H_2^+. In contrast, the second of (6.8) indicates that a repulsive force sets in upon the approach of H and H^+ in the anti-symmetric state Φ_a.

The solid curves $E_\mathrm{s}(R)$ and $E_\mathrm{a}(R)$ in the regions of $E < 0$ and $E > 0$ in Fig. 6.2(a) are drawn according to (6.23) and (6.27) in Sect. 6.3.3. In the calculations of (6.23) and (6.27) the terms representing the repulsion between the two protons, i.e., the fourth term in the square brackets of (6.1), is included so that both $E_\mathrm{s}(R)$ and $E_\mathrm{a}(R)$ represent the total energy of the H_2^+ system in the respective states.

According to Fig. 6.2(a), $E_\mathrm{s}(R)$ reaches a minimum at $R \approx 0.13\,\mathrm{nm}$, and the dissociation energy corresponding to the difference between this minimum and the dissociation limit is about $170\,\mathrm{kJ\,mol^{-1}} \cong 1.76\,\mathrm{eV}$. The broken line, on the other hand, is obtained by rigorous computations and precise experiments (cf. Sect. 6.4). The dissociation energy corresponding to the broken line is found to be $269\,\mathrm{kJ\,mol^{-1}} \cong 2.79\,\mathrm{eV}$ at $R \approx 0.106\,\mathrm{nm}$, which is about 1.6 times larger than the energy of 1.76 eV based on (6.27). This difference between the

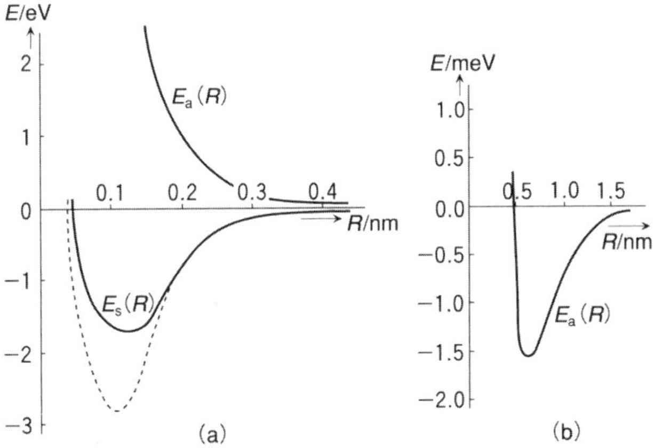

Fig. 6.2. (**a**): The total energies of H_2^+ in the symmetric ($E_s(R)$) and anti-symmetric ($E_a(R)$) states calculated by (6.27). The broken curve represents the results of rigorous calculations for the symmetric state, which is almost indistinguishable from experimentally determined curves. (**b**): The energy curve for the anti-symmetric state determined for extremely large values of R by a precise spectroscopic measurement. An explanation will be made in Sect. 6.4

approximate and the rigorous calculations should be regarded as tolerable because in the approximate calculation the curve for $E_s(R)$ is obtained by using the fixed 1s orbital of the H atom, which is not altogether realistic. The fair agreement between the two calculations strongly suggests that the use of the trial functions in (6.6) having the proper symmetry property are a shrewd thrust to the essence of chemical bond formation.

As to the apparently odd curve for Φ_a at $R > 0.5\,\text{nm}$ shown in Fig. 6.2(b), an explanation will be given in Sect. 6.4.

In concluding this section I will recapitulate the essential point of the chemical bond in the H_2^+ system because of its general importance. Firstly, the eigenstate of the H_2^+ ion cannot be described at any internuclear distance by either $\chi_l(\boldsymbol{r} + \boldsymbol{R}/2)$ nor $\chi_r(\boldsymbol{r} - \boldsymbol{R}/2)$ alone. Only after the superposition of them, i.e., after taking a linear combination of them, is an eigenfunction having the correct symmetry property obtained. Secondly, the symmetric Φ_s and the anti-symmetric Φ_a remain unchanged and change in sign, respectively, upon the exchange of $\boldsymbol{r} \leftrightarrow -\boldsymbol{r}$, or the exchange of $\chi_l(\boldsymbol{r}+\boldsymbol{R}/2) \leftrightarrow \chi_r(\boldsymbol{r}-\boldsymbol{R}/2)$. In the state Φ_s the behavior of the electron is always symmetric whatever the internuclear distance is, and the energy first decreases upon the approach of H and H^+ from infinity, passes a minimum, and then increases rapidly as the distance H and H^+ tends to 0. Thus, the system H_2^+ having only one electron can exist stably at the minimum. In contrast, in the state Φ_a, the energy is larger than the energy of an isolated atom and increases all the way with the approach of H and H^+.

In the next section we will see that the electron shuttles between the two protons when the symmetric and anti-symmetric states are superposed. For this purpose we will look into the motion of the wave packet of the electron.

6.1.2 The Motion of the Wave Packet of the Electron in H_2^+

Let us attach the time dependence to the eigenfunctions Φ_s and Φ_a in the preceding section to get the corresponding wavefunctions as follows (cf. the procedure in Sect. 2.1.2).

$$\begin{aligned} \Psi_s(\boldsymbol{r}, \boldsymbol{R}, t) &= \exp(-iE_s t/\hbar)\Phi_s(\boldsymbol{r}, \boldsymbol{R}) \\ \Psi_a(\boldsymbol{r}, \boldsymbol{R}, t) &= \exp(-iE_a t/\hbar)\Phi_a(\boldsymbol{r}, \boldsymbol{R}) \end{aligned} \tag{6.10}$$

According to the statement in connection with (2.10) and (2.11) in Sect. 2.1.4, the sum of $\Psi_s(\boldsymbol{r}, \boldsymbol{R}, t)$ and $\Psi_a(\boldsymbol{r}, \boldsymbol{R}, t)$ in (6.10) with a weight of, e.g., 1:1 as given in (6.11) is certainly a solution of the time-dependent Schrödinger equation, and therefore this is a possible state of our mechanical system.

$$\Psi_+(\boldsymbol{r}, \boldsymbol{R}, t) = \exp(-iE_s t/\hbar)\Phi_s(\boldsymbol{r}, \boldsymbol{R}) + \exp(-iE_a t/\hbar)\Phi_a(\boldsymbol{r}, \boldsymbol{R}) \tag{6.11}$$

$$i\hbar\frac{\partial}{\partial t}\Psi_+(\boldsymbol{r}, \boldsymbol{R}, t) = \hat{H}\Psi_+(\boldsymbol{r}, \boldsymbol{R}, t) \tag{6.12}$$

The two wave functions in (6.10) are to describe two quantum mechanical waves so that the function in (6.11) is regarded as describing a state of superposed waves in quantum mechanics. In (6.11), the plus combination was adopted, but the argument which follows is essentially the same if the minus combination is chosen instead.

Having the superposed state we next try to get the motion of the wave packet, which is described as proportional to the superposed state in (6.11) squared. From (6.11) we obtain the result below.

$$\begin{aligned} |\Psi_+(\boldsymbol{r}, \boldsymbol{R}, t)|^2 = |\Phi_s(\boldsymbol{r}, \boldsymbol{R})|^2 &+ |\Phi_a(\boldsymbol{r}, \boldsymbol{R})|^2 + \Phi_s^*(\boldsymbol{r}, \boldsymbol{R})\Phi_a(\boldsymbol{r}, \boldsymbol{R}) \\ &\times \exp\{-i(E_a - E_s)t/\hbar\} \\ &+\Phi_s(\boldsymbol{r}, \boldsymbol{R})\Phi_a^*(\boldsymbol{r}, \boldsymbol{R}) \exp\{+i(E_a - E_s)t/\hbar\} \end{aligned} \tag{6.13}$$

Since Φ_s and Φ_a are real, the above is rewritten as below in reference to the Euler equation in Sect. 2.2.3.

$$\begin{aligned} |\Psi_+(\boldsymbol{r}, \boldsymbol{R}, t)|^2 = |\Phi_s(\boldsymbol{r}, \boldsymbol{R})|^2 &+ |\Phi_a(\boldsymbol{r}, \boldsymbol{R})|^2 \\ &+2\Phi_s(\boldsymbol{r}, \boldsymbol{R})\Phi_a(\boldsymbol{r}, \boldsymbol{R}) \cos\{(E_a - E_s)t/\hbar\} \end{aligned} \tag{6.14}$$

When both $\Phi_s(\boldsymbol{r}, \boldsymbol{R})$ and $\Phi_a(\boldsymbol{r}, \boldsymbol{R})$ are substituted by the relations in (6.6) in the preceding Sect. 6.1.1, the following result is obtained.

$$
\begin{aligned}
|\Psi_+(\boldsymbol{r},\boldsymbol{R},t)|^2 &= \left[\frac{1}{1-S^2} + \frac{1}{\sqrt{1-S^2}}\cos\left\{\frac{(E_\mathrm{a}-E_\mathrm{s})t}{\hbar}\right\}\right]\chi_l^2(\boldsymbol{r}+\boldsymbol{R}/2) \\
&+ \left[\frac{1}{1-S^2} - \frac{1}{\sqrt{1-S^2}}\cos\left\{\frac{(E_\mathrm{a}-E_\mathrm{s})t}{\hbar}\right\}\right]\chi_r^2(\boldsymbol{r}-\boldsymbol{R}/2) \\
&- \frac{2S\chi_l(\boldsymbol{r}+\boldsymbol{R}/2)\chi_r(\boldsymbol{r}-\boldsymbol{R}/2)}{1-S^2}
\end{aligned} \tag{6.15}
$$

In the right-hand side of the above equation the time-independent parts are combined as, $\chi_l^2(\boldsymbol{r}+\boldsymbol{R}/2)/(1-S^2)+\chi_r^2(\boldsymbol{r}-\boldsymbol{R}/2)/(1-S^2)-2S\chi_l(\boldsymbol{r}+\boldsymbol{R}/2)\times\chi_r(\boldsymbol{r}-\boldsymbol{R}/2)/(1-S^2)$ which can be approximated as $\chi_l^2(\boldsymbol{r}+\boldsymbol{R}/2)+\chi_r^2(\boldsymbol{r}-\boldsymbol{R}/2)$ when the overlap integral S is small. This result simply means that in the 1:1 superposed state, the time-independent part of the probability of finding the electron is centered on both protons.

What is interesting to us is the time-dependent part in (6.15), which is extracted as below.

$$
\begin{aligned}
&\left[(1/\sqrt{1-S^2})\cos\{(E_\mathrm{a}-E_\mathrm{s})t/\hbar\}\right]\chi_l^2(\boldsymbol{r}+\boldsymbol{R}/2) \\
&-\left[(1/\sqrt{1-S^2})\cos\{(E_\mathrm{a}-E_\mathrm{s})t/\hbar\}\right]\chi_r^2(\boldsymbol{r}-\boldsymbol{R}/2)
\end{aligned} \tag{6.16}
$$

As the value of the cosine argument, $(E_\mathrm{a}-E_\mathrm{s})t/\hbar \equiv 2\pi\Delta Et/h$, increases from 0 as $0, \pi/2, \pi, 3\pi/2, 2\pi, \ldots$, the value of the cosine changes as $1, 0, -1, 0, 1, \ldots$. Therefore, the time-dependent part in the superposed state of $\Psi_+(\boldsymbol{r},\boldsymbol{R},t)$ undulates sinusoidally as shown in Fig. 6.3. A similar but simpler example has been demonstrated in Sect. 2.2.3.

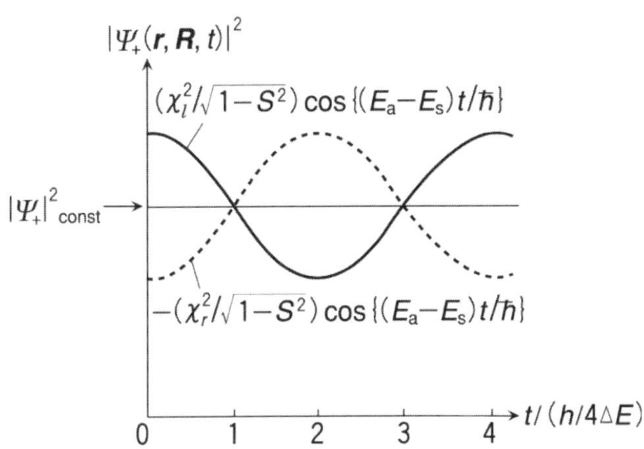

Fig. 6.3. The time dependence of the three components of $|\Psi_+(\boldsymbol{r},\boldsymbol{R},t)|^2$ in (6.15). The solid and the dotted sinusoidal curves represent the contributions from the left and the right protons, respectively. The horizontal line corresponds to the time-independent contribution in (6.15)

In Fig. 6.3 the horizontal line shows the time-independent part $|\Psi_+|^2_{\rm const} \equiv \chi_l^2(\boldsymbol{r}+\boldsymbol{R}/2)/(1-S^2)+\chi_r^2(\boldsymbol{r}-\boldsymbol{R}/2)/(1-S^2)-2S\chi_l(\boldsymbol{r}+\boldsymbol{R}/2)\times\chi_r(\boldsymbol{r}-\boldsymbol{R}/2)/(1-S^2)$, while the two curves demonstrate the temporal change in $|\Psi_+(\boldsymbol{r},\boldsymbol{R},t)|^2$.

From the figure it is seen that if the electron is localized initially near the proton at the left, then it starts moving toward the right with time and repeats in periodical motion.

According to the light and easy phrasing from Tomonaga's book cited at the beginning of this section, the above situation is described as, "We therefore can say the exchange force originates in the shuttling of the electron; if we express this shuttling of the electron in other words, we may also imagine that nucleus I (left) and nucleus II (right) alternately don the electron. Or we can say that the particle at $-\boldsymbol{R}/2$ and the particle at $\boldsymbol{R}/2$ are alternately becoming the neutral atom and the positive ion. This *alteration* is the essence of the exchange force" (quoted from Oka's translation cited above).

The frequency of the periodic motion of the wave packet shown in Fig. 6.3 is given by the relation $h\nu = E_{\rm a} - E_{\rm s} = \Delta E$. Since the energy difference near the equilibrium internuclear distance is large compared with that for large internuclear distances as shown in Fig. 6.2, the shuttling of the electron is frequent near the equilibrium internuclear distance. At infinite internuclear distance the frequency reduces to 0, which means that the electron stays localized permanently around one of the two protons.

If the H_2^+ system in the symmetric ground state is irradiated with an electromagnetic wave with a frequency given by the above equation, the electron in the H_2^+ system will be resonantly excited keeping the spin state unchanged. Here, the resonance means that the system repeats the absorption and emission of the photon at an energy of $h\nu = E_{\rm a} - E_{\rm s}$, i.e., with the same frequency of the electron shuttling between the two protons. Since the excited state is of a dissociative character, however, the resonating vertical (optical) transition between the two states cannot last forever but must compete with the dissociation in the excited state. The vertical transition is, in fact, observed in the ultraviolet (UV) region at energies of about 10 to 15 eV. The transition may be called the 'charge resonance transition', because it is inherently associated with the electron shuttling which occurs in resonance with the optical transition.

The time needed for the electron to perform one cycle of the shuttling is calculated by setting the cosine argument in (6.16), $(E_{\rm a} - E_{\rm s})t/\hbar$, equal to 2π. Assuming that the energy difference $E_{\rm a} - E_{\rm s} = \Delta E$ is roughly 10 eV for simplicity, the time for one cycle is estimated as $t \approx 4\times10^{-16} = 0.4$ fs. Thus, one could monitor the electron shuttling between the protons upon the superposition of the symmetric and anti-symmetric states, provided that the time resolution of the monitoring of the system is high (fast) enough. In actuality, the motion is too fast to follow by present techniques. However, the wave packet motion of some nuclei, whose masses are more than four orders of magnitude larger than the mass of the electron, is actually observed because the time of motion for these heavy nuclei falls in the ps (10^{-12} s) regime which is experimentally accessible (cf. the nuclear wave packet motion of NaI in Sect. 8.3.1).

It bears keeping in mind that the two wavefunctions in (6.10) employed for the superposition represent stationary states as explained in Sect. 2.1.2. Therefore, the energy eigenvalues E_s and E_a remain constant with time, of course. These stationary states are recovered from the dynamically changing $\Psi_+(\boldsymbol{r}, \boldsymbol{R}, t)$ and $\Psi_-(\boldsymbol{r}, \boldsymbol{R}, t)$ by proper linear combinations of the latter two (cf. the argument in connection with (6.11) and (6.12) for $\Psi_+(\boldsymbol{r}, \boldsymbol{R}, t)$ and $\Psi_-(\boldsymbol{r}, \boldsymbol{R}, t)$).

In passing, the above energy difference $\Delta E = E_a - E_s$ will be shown to be approximately equal to an energy of $2K$ where K is called the exchange integral. This integral, appearing in (6.24) in Sect. 6.3.3, has a fundamental importance in stabilizing the H_2^+ system in the ground state (cf. Sect. 6.3.3).

I close this rather lengthy section by citing once again Tomonaga, "We therefore can say the exchange force originates in the shuttling of the electron, ...". With this citation I hope that the essential stability of the two protons-one electron system is understood in such terms as the exchange force and the superposition of states.

6.2 The Stability of the H_2^+ Ion (Explanation 2): Qualitative Explanation in Terms of the Uncertainty Principle

As shown in (6.8) and in Fig. 6.2(a), the eigenvalue of energy of the H_2^+ ion corresponding to the symmetric eigenfunction Φ_s is lower than the energy of the isolated H atom E_{1s}, while the eigenvalue for the anti-symmetric eigenfunction Φ_a is higher than E_{1s}. The respective stability and instability of the states Φ_s and Φ_a can be qualitatively explained, after R.P. Feynman, by referring to the uncertainty principle as follows.

Consider the ground electronic state of an isolated H atom. The electron will tend to approach the proton as much as possible to lower the average potential energy. As the electron is confined in a small space, the fluctuation of the positional coordinate Δx of the electron is also reduced. Accordingly, the fluctuation of the momentum Δp must increase by the relation of $\Delta x \Delta p \geq \hbar/2$ given in Sect. 1.2, which entails the increase in the average kinetic energy of the electron. The stationary total energy of E_{1s} is just the sum of the two counterbalancing contributions.

When the system is switched from H to H_2^+, the single electron finds another proton and the spatial region of lower potential energies is expanded in the symmetric state Φ_s. The electron can now move in a relatively gentle manner in this region with a larger Δx and so with a smaller Δp. As a result, the kinetic energy is lowered relative to the case in an isolated H atom to attain a total energy that is smaller than E_{1s}. Contrastingly, in the anti-symmetric state $\Phi_a = [\chi_l(\boldsymbol{r} + \boldsymbol{R}/2) - \chi_r(\boldsymbol{r} - \boldsymbol{R}/2)] / \sqrt{2(1-S)}$ (cf. (6.6)), the electron will see one node (cf. Sect. 3.2.5) at the midpoint of the two protons where the probability of finding the electron becomes 0 (cf. Sect. 6.3.1). Under such a circumstance, the electron is again confined to one of the separated spatial regions with a smaller Δx and a higher Δp, resulting in a total energy of E_a which is higher than E_{1s}.

6.3 The Stability of the H_2^+ Ion (Explanation 3): An Explanation in Terms of the LCAO-MO Method

The abbreviation LCAO comes from the phrase 'Linear Combination of Atomic orbital's. It is the simplest approximation to describe the electron in molecules in terms of atomic orbitals. The linear combination is called the Molecular Orbital (MO), a natural extension of the concept of the Atomic Orbital (AO) in Chap. 3. The physical justification of the approximation is based on the assumption that electrons may behave similarly to that in isolated atoms when they approach the nuclei in molecules.

The contributors to the LCAO theory in the early days of quantum chemistry include R.S. Mulliken, F. Hund, E.A.J. Hückel, J.C. Slater and others. The theory will be treated in detail in Chaps. 7 and 8. In this chapter, the theory will be examined to aid in the understanding of the essence of the bonding in the simplest molecule H_2^+.

As a matter of fact, we have already been exposed to LCAO-MO's; the two functions Φ_s and Φ_a in (6.6) are linear combinations of the two AO's of the H atom. It is to be noted, however, that (6.6) was derived from the removal of the exchange degeneracy, whereas the essence of the above stated LCAO-MO theory is based on chemists' intuition.

As mentioned in Sect. 6.1, however, the LCAO-MO has an intrinsic fault in that atoms in molecules cannot remain the same as isolated atoms so that the use of the fixed AO in the molecular orbital has obvious limitations. However, in compensation for this drawback, simple linear combinations of AO allow for easy computational manipulations. Moreover, the inflexible AO can be replaced, on demand, with various adjustable functions called, in general, basis functions. As will be shown in Sect. 8.2.1, nowadays a number of useful basis functions are included in computational program packages. In this section however, I will concentrate on the primitive LCAO-MO for the system. In the following Sects. 6.3.1 and 6.3.2 I will introduce two methods of obtaining the LCAO-MO for the system. This procedure may be redundant but the two methods each have their own significance. In the last Sect. 6.3.3, a guide to compute integrals necessary for obtaining the energy eigenvalues will be given.

6.3.1 Determination of the LCAO-MO of H_2^+ by the Use of Symmetry

First, the eigenvalue equation for energy of the system is reproduced below (cf. Fig. 6.1).

$$\begin{aligned}&\hat{H}(\boldsymbol{r},\boldsymbol{R})\Phi(\boldsymbol{r},\boldsymbol{R})\\&=\left[\frac{-\hbar^2}{2m_\mathrm{e}}\Delta-\frac{e^2}{4\pi\varepsilon_0|\boldsymbol{r}+\boldsymbol{R}/2|}-\frac{e^2}{4\pi\varepsilon_0|\boldsymbol{r}-\boldsymbol{R}/2|}+\frac{e^2}{4\pi\varepsilon_0 R}\right]\Phi(\boldsymbol{r},\boldsymbol{R})\\&=E\Phi(\boldsymbol{r},\boldsymbol{R})\end{aligned} \tag{6.1}$$

Since the H_2^+ ion happens to be a one-electron system, the eigenfunction $\Phi(\boldsymbol{r},\boldsymbol{R})$ should be represented by some one-electron function $\varphi(\boldsymbol{r},\boldsymbol{R})$. As a trial, let us assume that this function can be approximated by the following linear combination of the known one-electron orbitals, $\chi_l(\boldsymbol{r}+\boldsymbol{R}/2)$ and $\chi_r(\boldsymbol{r}-\boldsymbol{R}/2)$:

$$\varphi(\boldsymbol{r},\boldsymbol{R})=c_l\chi_l(\boldsymbol{r}+\boldsymbol{R}/2)+c_r\chi_r(\boldsymbol{r}-\boldsymbol{R}/2) \tag{6.17}$$

Then, within this framework, the best coefficients c_l and c_r should lead to the best $\varphi(\boldsymbol{r},\boldsymbol{R})$ and thus, the best approximation to $\Phi(\boldsymbol{r},\boldsymbol{R})$.

In H_2^+, the contributions of $\chi_l(\boldsymbol{r}+\boldsymbol{R}/2)$ and $\chi_r(\boldsymbol{r}-\boldsymbol{R}/2)$ should be equal by symmetry, which requires the equality of $c_l^2=c_r^2$, so the relation $c_l=\pm c_r$, and so $\varphi(\boldsymbol{r},\boldsymbol{R})=c_l\{\chi_l(\boldsymbol{r}+\boldsymbol{R}/2)\pm\chi_r(\boldsymbol{r}-\boldsymbol{R}/2)\}$. By asking this function to be normalized, we have the following relation.

$$\begin{aligned}&\int|\varphi(\boldsymbol{r},\boldsymbol{R})|^2\mathrm{d}v\\&=c_l^2\left[\int\left\{\chi_l^2(\boldsymbol{r}+\boldsymbol{R}/2)\pm2\chi_l(\boldsymbol{r}+\boldsymbol{R}/2)\chi_r(\boldsymbol{r}-\boldsymbol{R}/2)+\chi_r^2(\boldsymbol{r}-\boldsymbol{R}/2)\right\}\mathrm{d}v\right]\\&=1\end{aligned} \tag{6.18}$$

Since both $\chi_l(\boldsymbol{r}+\boldsymbol{R}/2)$ and $\chi_r(\boldsymbol{r}-\boldsymbol{R}/2)$ are already normalized, and since the overlap integral $S=\int\chi_l(\boldsymbol{r}+\boldsymbol{R}/2)\chi_r(\boldsymbol{r}-\boldsymbol{R}/2)\mathrm{d}v$ can be obtained elementarily as given in (6.9), the coefficient c_l is now explicitly obtained as $c_l=1/\sqrt{2(1\pm S)}$. Altogether, the concrete forms of (6.17) are found to coincide with Φ_s and Φ_a in (6.6) below which have previously been obtained by the perturbation method (cf. Sect. 6.1.1).

$$\begin{aligned}\Phi_\mathrm{s}(\boldsymbol{r},\boldsymbol{R})\equiv\Phi_\mathrm{s}&=\left[\chi_l(\boldsymbol{r}+\boldsymbol{R}/2)+\chi_r(\boldsymbol{r}-\boldsymbol{R}/2)\right]/\sqrt{2(1+S)}\\\Phi_\mathrm{a}(\boldsymbol{r},\boldsymbol{R})\equiv\Phi_\mathrm{a}&=\left[\chi_l(\boldsymbol{r}+\boldsymbol{R}/2)-\chi_r(\boldsymbol{r}-\boldsymbol{R}/2)\right]/\sqrt{2(1-S)}\end{aligned} \tag{6.6}$$

These linear combinations of $\chi_l(\boldsymbol{r}+\boldsymbol{R}/2)$ and $\chi_r(\boldsymbol{r}-\boldsymbol{R}/2)$ can be illustrated in Fig. 6.4.

The probabilities of finding the electron in the orbitals φ_s and φ_a are given by $|\varphi_\mathrm{s}|^2$ and $|\varphi_\mathrm{a}|^2$, respectively. As is seen from (6.6), the normalization

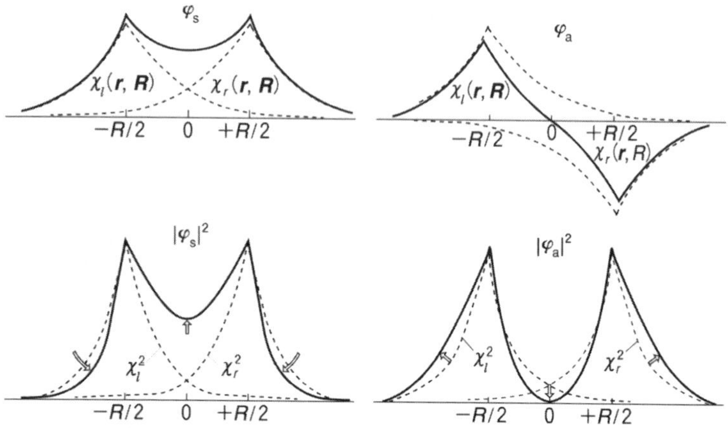

Fig. 6.4. Symmetric MO φ_s and anti-symmetric MO φ_a of H_2^+ (upper), and their squared functions (lower)

constant for φ_s is smaller than $1/\sqrt{2}$. Therefore, the probability is reduced at distances away from the protons and enhanced between them in comparison with the 1s orbital of the isolated H atom, as indicated by the arrows for $|\varphi_s|^2$ in Fig. 6.4. Conversely, the normalization constant for φ_a is larger than $1/\sqrt{2}$, causing the expansion of the electron distribution outward at the sacrifice of the electron distribution at $R = 0$ where the following equation holds,

$$|\varphi_a(\boldsymbol{r}, 0)|^2 = \left[\chi_l^2(\boldsymbol{r}, 0) - 2\chi_l(\boldsymbol{r}, 0)\chi_r(\boldsymbol{r}, 0) + \chi_r^2(\boldsymbol{r}, 0)\right] / [2(1 - S)] = 0 \, .$$

Therefore, the electron in φ_s tends to be localized in the region between the protons, preventing them from repelling each other, while the electron in φ_a diffuses away from the molecular center, increasing the repulsion between the protons. In this sense, it is reasonable to call φ_s and φ_a the bonding orbital and the anti-bonding orbital, respectively.

Once the eigenfunction $\Phi(\boldsymbol{r}, \boldsymbol{R})$ in (6.1) is approximated by φ_s and φ_a, the next task is to compute the eigenvalues of energy $E_{s,a}$ by the equation, $E_{s,a} = \int \Phi_{s,a}^* \hat{H} \Phi_{s,a} dv = \int \varphi_{s,a}^* \hat{H} \varphi_{s,a} dv$. However, this will be postponed until Sect. 6.3.3, and in next, in Sect. 6.3.2, it will be demonstrated that the result of (6.6) is also attained by the method of Ritz's calculus of variation explained in Sect. 2.1.15.

6.3.2 Determination of the LCAO-MO of H_2^+ by Ritz's Calculus of Variation

In the preceding section, the LCAO-MO of H_2^+ was determined by utilizing the symmetry of the system. In this section we search for the MO through the method of Ritz's calculus of variation, which can be applied to systems having no specific symmetry. As in Sect. 6.3.1, we start from the eigenvalue equation (6.1) and the trial function reproduced below.

$$\varphi(\boldsymbol{r},\boldsymbol{R}) = c_l\chi_l(\boldsymbol{r}+\boldsymbol{R}/2) + c_r\chi_r(\boldsymbol{r}-\boldsymbol{R}/2) \tag{6.17}$$

The functions $\chi_l(\boldsymbol{r}+\boldsymbol{R}/2)$ and $\chi_r(\boldsymbol{r}-\boldsymbol{R}/2)$ in (6.17) correspond to the general function $\chi_i(\boldsymbol{r})$ in the trial function $\varphi(\boldsymbol{r}) = \sum_i c_i\chi_i(\boldsymbol{r})$ in Sect. 2.1.15. (Nuclear coordinates such as R in (6.17) are regarded as parameters and omitted in $\varphi(\boldsymbol{r})$ and $\chi_i(\boldsymbol{r})$). In the following we will try to determine c_l and c_r in (6.17) by the procedures of (2.57) through (2.61) in Sect. 2.1.15. The simultaneous equation in (2.59) to obtain the linear coefficients becomes in the present case,

$$\begin{aligned}(H_{ll}-\varepsilon S_{ll})c_l + (H_{lr}-\varepsilon S_{lr})c_r &= 0\\ (H_{rl}-\varepsilon S_{rl})c_l + (H_{rr}-\varepsilon S_{rr})c_r &= 0\,.\end{aligned} \tag{6.19}$$

Therefore, the secular equation corresponding to (2.61) is,

$$\begin{vmatrix} H_{ll}-\varepsilon S_{ll} & H_{lr}-\varepsilon S_{lr}\\ H_{rl}-\varepsilon S_{rl} & H_{rr}-\varepsilon S_{rr}\end{vmatrix} = 0\,. \tag{6.20}$$

Differing from Sect. 6.3.1, where use was made of the symmetry of the system, we now need the values of the finite integrals H_{ll}, $H_{lr} = H_{rl}$, H_{rr}, and $S_{lr} = S_{rl}$ (both S_{ll} and S_{rr} can be put to 1 because the atomic orbitals $\chi_l(\boldsymbol{r}+\boldsymbol{R}/2)$ and $\chi_r(\boldsymbol{r}-\boldsymbol{R}/2)$ are assumed to have been normalized). Once these integrals are obtained, we can find the two roots of ε and the corresponding pair of coefficients (c_{1l}, c_{1r}) and (c_{2l}, c_{2r}), via Ritz's calculus of variation as delineated in Sect. 2.1.15.

The needed integrals can be computed elementarily as given in the next Sect. 6.3.3. However, in the present, simple case of H_2^+, the following relations $H_{ll} = H_{rr}$, $H_{lr} = H_{rl}$, and $S_{lr} = S_{rl}$, hold obviously. Therefore, if these integrals are abbreviated as α, β, and S, respectively, the secular equation in (6.20) can be rewritten as in (6.21).

$$\begin{vmatrix} \alpha-\varepsilon & \beta-\varepsilon S\\ \beta-\varepsilon S & \alpha-\varepsilon\end{vmatrix} = 0 \tag{6.21}$$

The two eigenvalues are obtained immediately as below.

$$E_\mathrm{s} = (\alpha+\beta)/(1+S), \quad E_\mathrm{a} = (\alpha-\beta)/(1-S) \tag{6.22}$$

Substituting these eigenvalues into (6.19), the corresponding bonding orbital φ_s and anti-bonding φ_a, are obtained which, of course, coincide with the previously obtained results in (6.6).

$$\begin{aligned}\varphi_\mathrm{s} &= [\chi_l(\boldsymbol{r}+\boldsymbol{R}/2) + \chi_r(\boldsymbol{r}-\boldsymbol{R}/2)]\,/\sqrt{2(1+S)}\\ \varphi_\mathrm{a} &= [\chi_l(\boldsymbol{r}+\boldsymbol{R}/2) - \chi_r(\boldsymbol{r}-\boldsymbol{R}/2)]\,/\sqrt{2(1-S)}\end{aligned} \tag{6.6}$$

Since the integral α in (6.22) is already known as a negative quantity and S is subject to the relationship $0 < S < 1$, it is concluded that β must be a negative quantity from the relation $E_\mathrm{a} - E_\mathrm{s} = 2(\alpha S - \beta)/(1-S^2) > 0$.

6.3.3 Coulomb Integral and the Exchange Integral of the H_2^+ System

The two energy eigenvalues of interest, E_{s} and E_{a}, are given below in reference to (6.1).

$$
\begin{aligned}
E_{\mathrm{s,a}} &= \int \varphi^*_{\mathrm{s,a}} \hat{H} \varphi_{\mathrm{s,a}} \mathrm{d}v \\
&= \frac{1}{2(1 \pm S)} \int [\chi_l^*(\boldsymbol{r} + \boldsymbol{R}/2) \pm \chi_r^*(\boldsymbol{r} - \boldsymbol{R}/2)] \\
&\quad \times \left[\frac{-\hbar^2}{2m_{\mathrm{e}}} \Delta - \frac{e^2}{4\pi\varepsilon_0} \left(\frac{1}{|\boldsymbol{r} + \boldsymbol{R}/2|} + \frac{1}{|\boldsymbol{r} - \boldsymbol{R}/2|} - \frac{1}{R} \right) \right] \\
&\quad \times [\chi_l(\boldsymbol{r} + \boldsymbol{R}/2) \pm \chi_r(\boldsymbol{r} - \boldsymbol{R}/2)] \, \mathrm{d}v
\end{aligned}
\tag{6.23}
$$

The two functions $\chi_l(\boldsymbol{r} + \boldsymbol{R}/2)$ and $\chi_r(\boldsymbol{r} - \boldsymbol{R}/2)$ are equal to the 1s atomic orbitals of the H atom $1\mathrm{s}(|\boldsymbol{r} + \boldsymbol{R}/2|)$ and $1\mathrm{s}(|\boldsymbol{r} - \boldsymbol{R}/2|)$, respectively (cf. Fig. 6.1). Therefore, it might be expected that the integrals appearing in (6.19) through (6.23) could be obtained easily in the polar coordinate systems centered on the two protons. This is not the case, however, because the integral of the single electron over the whole space must be performed in one common coordinate system. This is another example of the importance of choosing a proper coordinate system to obtain analytic solutions of the problem under consideration (cf. Sect. 3.2.1).

The proper coordinates in the present system are the confocal elliptic coordinates in Fig. 6.5. In this coordinate system, the whole space is covered by ellipses $1, 2, 3, \ldots$ and parabolas $1, 2, 3, \ldots$ orthogonal to the respective ellipses in the xy-plane and by the rotation of both types of the curves around the x-axis over 0 through 2π. The ellipses and the parabolas share the common focuses $(F, 0, 0)$ and $(-F, 0, 0)$. Any position in the three-dimensional space is fixed by the set of r_{A}, r_{B}, and φ, or the set of λ, μ, and φ as shown in Fig. 6.5.

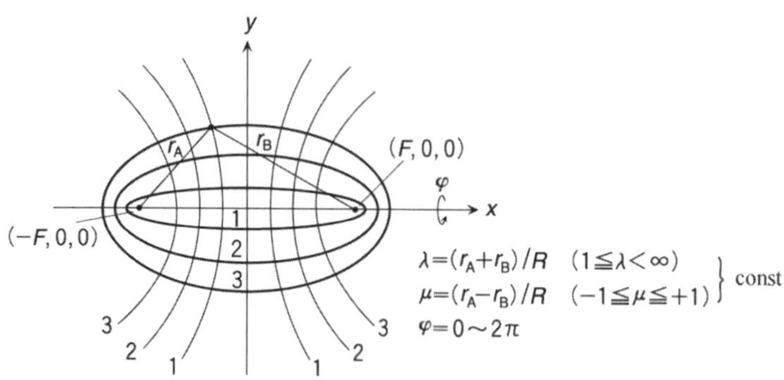

Fig. 6.5. Confocal elliptic coordinate system

By expanding the integral in (6.23), several terms appear; it is instructive to examine the physical meaning of these terms.

Some of the terms can be associated with what should by now be familiar physical quantities. For example, if we look at the term, $\int \chi_l^*(\boldsymbol{r} + \boldsymbol{R}/2)\left[-\hbar^2 \Delta/(2m_\mathrm{e}) - e^2/(4\pi\varepsilon_0|\boldsymbol{r} + \boldsymbol{R}/2|)\right]\chi_l(\boldsymbol{r}+\boldsymbol{R}/2)\mathrm{d}v$, it represents nothing else but the ground state energy of the electron localized on the left-hand proton because of the obvious relationship, $\left[-\hbar^2\Delta/(2m_\mathrm{e}) - e^2/(4\pi\varepsilon_0|\boldsymbol{r} \pm \boldsymbol{R}/2|)\right]\chi_{l,r}(\boldsymbol{r} \pm \boldsymbol{R}/2) = E_{1\mathrm{s}}\chi_{l,r}(\boldsymbol{r} \pm \boldsymbol{R}/2)$. Here, the plus and the minus signs correspond to the subscripts l and r, respectively.

Paying attention to these relationships and continuing the rearrangement of the expanded terms, we finally reach the following concise expression for (6.23).

$$E_{\mathrm{s,a}} = E_{1\mathrm{s}} + \frac{J}{1 \pm S} \pm \frac{K}{1 \pm S} \tag{6.24}$$

The symbols J and K in (6.24) stand for the two integrals in (6.25) and (6.26) below.

$$\begin{aligned} J &\equiv \frac{e^2}{4\pi\varepsilon_0}\left[-\int \frac{\chi_l^*(\boldsymbol{r}+\boldsymbol{R}/2)\chi_l(\boldsymbol{r}+\boldsymbol{R}/2)}{|\boldsymbol{r}-\boldsymbol{R}/2|}\mathrm{d}v + \frac{1}{R}\right] \\ &= \frac{e^2}{4\pi\varepsilon_0}\left[-\int \frac{\chi_r^*(\boldsymbol{r}-\boldsymbol{R}/2)\chi_r(\boldsymbol{r}-\boldsymbol{R}/2)}{|\boldsymbol{r}+\boldsymbol{R}/2|}\mathrm{d}v + \frac{1}{R}\right] \end{aligned} \tag{6.25}$$

and

$$\begin{aligned} K &\equiv \frac{e^2}{4\pi\varepsilon_0}\left[-\int \frac{\chi_l^*(\boldsymbol{r}+\boldsymbol{R}/2)\chi_r(\boldsymbol{r}-\boldsymbol{R}/2)}{|\boldsymbol{r}+\boldsymbol{R}/2|}\mathrm{d}v \right. \\ &\quad \left. + \frac{1}{R}\int \chi_l^*(\boldsymbol{r}+\boldsymbol{R}/2)\chi_r(\boldsymbol{r}-\boldsymbol{R}/2)\mathrm{d}v\right] \\ &= \frac{e^2}{4\pi\varepsilon_0}\left[-\int \frac{\chi_r^*(\boldsymbol{r}-\boldsymbol{R}/2)\chi_l(\boldsymbol{r}+\boldsymbol{R}/2)}{|\boldsymbol{r}-\boldsymbol{R}/2|}\mathrm{d}v \right. \\ &\quad \left. + \frac{1}{R}\int \chi_r^*(\boldsymbol{r}-\boldsymbol{R}/2)\chi_l(\boldsymbol{r}+\boldsymbol{R}/2)\mathrm{d}v\right] \end{aligned} \tag{6.26}$$

The quantities J and K are named the Coulomb integral and the exchange integral, respectively. Note that the same symbols with the same names will appear in Sect. 7.2.1 (cf. (7.12)) in the argument of the energy of the H_2 molecule. However, they are different from those in (6.25) and (6.26) above.

It is seen from (6.25) that the first term in the square brackets represents the energy of the Coulombic attractive interaction between the electron densities localized on the left or the right proton, i.e., $\chi_l^*(\boldsymbol{r}+\boldsymbol{R}/2)\chi_l(\boldsymbol{r}+\boldsymbol{R}/2)$, or $\chi_r^*(\boldsymbol{r}-\boldsymbol{R}/2)\chi_r(\boldsymbol{r}-\boldsymbol{R}/2)$ with the counterpart protons at the right or the left, respectively. Thus, the name of the Coulomb integral for J is justifiable. On the other hand, the two integrals in the brackets for K in (6.26) imply that both protons have some relation with the electron. The origin of these two integrals can be traced back to the fact that the eigenfunctions in (6.6) are described by the superposition of the two states $\chi_l(\boldsymbol{r}+\boldsymbol{R}/2)$ and $\chi_r(\boldsymbol{r}-\boldsymbol{R}/2)$. Although the exchange integral K cannot be associated with a classical interpretation contrary to J, it has deep significance in the world dictated by quantum mechanics.

Up to (6.24) through (6.26), we have merely rearranged the expanded terms of (6.23). In order to get $E_{\rm s,a}$ in (6.23), we have to perform the computation in the confocal elliptic coordinate system as forecasted above. Fortunately, the actual calculation can be carried out analytically in terms of elementary integrals. Here, only the result is given in (6.27) below where a_0 is the Bohr radius. The overlap integral S is already explicitly given in (6.9). The sign $\pm$ correspond to $E_{\rm s}$ $(+)$ and $E_{\rm a}$ $(-)$.

$$\begin{aligned} E_{\rm s,a} &= E_{\rm 1s} + \frac{e^2}{4\pi\varepsilon_0 a_0(1\pm S)}\left(1+\frac{a_0}{R}\right){\rm e}^{-2R/a_0} \\ &\pm \frac{e^2}{4\pi\varepsilon_0(1\pm S)}\left\{\frac{S}{R}-\frac{1}{a_0}\left(1+\frac{R}{a_0}\right){\rm e}^{-R/a_0}\right\} \end{aligned} \tag{6.27}$$

The first through third terms of the right-hand side of (6.27) correspond, respectively, to the first through third terms of (6.24). As is seen from (6.27), the energy of H_2^+ is always destabilized by the second term, i.e., by the Coulombic interaction energy, relative to the energy $E_{\rm 1s}$ of the H atom in the ground electronic state while the third term, i.e., the exchange interaction energy, causes the main splitting into $E_{\rm s}$ and $E_{\rm a}$. The solid curves in Fig. 6.2(a) are obtained according to (6.27). As mentioned in Sect. 6.1.1, (6.27) approaches (6.8) at large internuclear distances.

6.4 A Few Topics of Study in the H_2^+ System

As mentioned in the preceding section, the H_2^+ system is unique in the sense that it is the simplest molecule having uniaxial symmetry and a single electron so that both the eigenfunction and the eigenvalue of energy are analytically obtainable. This situation has attracted the interest of a number of theorists since 1927 up to the present. When a new molecular theory is developed, this simplest system is studied as a testbed. In recent years, theories beyond the Born-Oppenheimer approximation are being developed for the H_2^+ system. However, it is not our purpose to command the frontier in the study of molecular systems. Instead, I should like to emphasize that the system is most important in the fundamental understanding of chemical bonds even at an elementary level.

The H_2^+ system has been energetically studied through experiment also (e.g., J. Posthumus, 2001). As a result of spectroscopic studies, it has been revealed that there exists a very shallow potential minimum with a depth of about 1.5 meV at an internuclear distance of about 0.6 nm for the state approximated by $\varphi_a(\boldsymbol{r}, \boldsymbol{R})$ (or $\Phi_a(\boldsymbol{r}, \boldsymbol{R})$), which has been regarded as a state of anti-bonding character. The small minimum is amplified in Fig. 6.2(b) (A. Carrington, 1995).

This experimental finding is not contradictory to what has been stated up to now because in the actual H_2^+ system, the use of the inflexible 1s orbital of the H atom as a basis function cannot be said to be reality. The electron approximated by the 1s orbital must be influenced by the positive charge of the counterpart proton giving rise to additional stabilization. Thus, the H_2^+ ion at the shallow minimum may be regarded as a sort of the van der Waals complex.

7 The Hydrogen Molecule: Why are Two Neutral Hydrogen Atoms Stabilized by Simply Approaching Each Other?

There are a number of stable diatomic molecules in nature which consist of two neutral atoms. In these molecules, electrons do not just cancel out the repulsion between the two positively charged bare nuclei but contribute significantly to the stabilization of the molecules as a whole. In this chapter we investigate the stabilization mechanism by analyzing the H_2 molecule as a simple system.

7.1 It is Spectroscopic Experiments on the H_2 Molecule that have Paved the Way to Quantum Chemical Theory

The Heitler-London theory, the very starting point of molecular quantum mechanics, would not have seen the light of day had it not been for the antecedent spectroscopic experiments. These experiments provided the data which, in the wisdom of hindsight, foreshadowed the birth of the theory.

7.2 The Heitler-London Theory

The initial attempt of Heitler and London was aimed at estimating quantitatively the van der Waals force between two H atoms, the concept of which had been familiar since the nineteenth century. For this purpose the two young men applied the newborn quantum mechanics, without knowing now familiar concepts and techniques such as the expectation value and the calculus of variation. In the end, they arrived successfully at the correct eigenfunction of the H_2 molecule by considering the indistinguishability of two H atoms and the orthogonality of the eigenfunction. The discovery of the correct eigenfunction having the proper symmetry was an important achievement in the understanding of a chemical bond between two neutral atoms.

7.3 Development of the Heitler-London Theory

The H_2 molecule is a basic system for both theoretical and experimental studies of molecules. A few topics concerning more sophisticated studies on the system will be briefly introduced.

7.4 Molecular Orbital Treatments of the H_2 Molecule

In view of the versatility of the molecular orbital theory a few comments will be made on the treatment of the H_2 molecule by the molecular orbital theory.

7.1 It is Spectroscopic Experiments on the H_2 Molecule that have Paved the Way to Quantum Chemical Theory

Before proceeding to the theory of the chemical bond in the H_2 molecule, I will digress a little by first providing relevant background concerning some experimental studies of the H_2 molecule. Remember that development of the Bohr theory for the H atom was spurred by the spectroscopic information provided by astronomers and spectroscopists (cf. Sect. 1.4). Just as this treasure of data was mined for the H atom, the birth of molecular quantum mechanics was also aided by preceding experimental information on the electronic spectrum of the H_2 molecule. In general, molecular spectra are much more complicated than atomic spectra, because in molecules not only the electronic energy state but also energy states due to the nuclear rotation and vibration are involved. Despite this complexity, researchers had been patiently accumulating molecular spectral data since the nineteenth century. This activity was indispensable for the birth of molecular quantum mechanics as will be explained.

Since the discovery by V. Schumann in the late nineteenth century that the H_2 molecule exhibits emissions in the ultraviolet region, numerous investigations on the optical transitions of the molecule were carried out. In 1925, T. Lyman reported that the transition of the molecule from the first excited electronic state to the ground electronic state can be understood in reference to the schematic energy diagram in Fig. 7.1(a), i.e., the emission spectrum consists of a series of transitions from the vibrational state $v' = 0$ in an excited electronic state to the states $v'' = 0, 1, 2, \ldots$ in the ground electronic state (cf. Fig. 5.1 for the quantized rotation-vibration state in general).

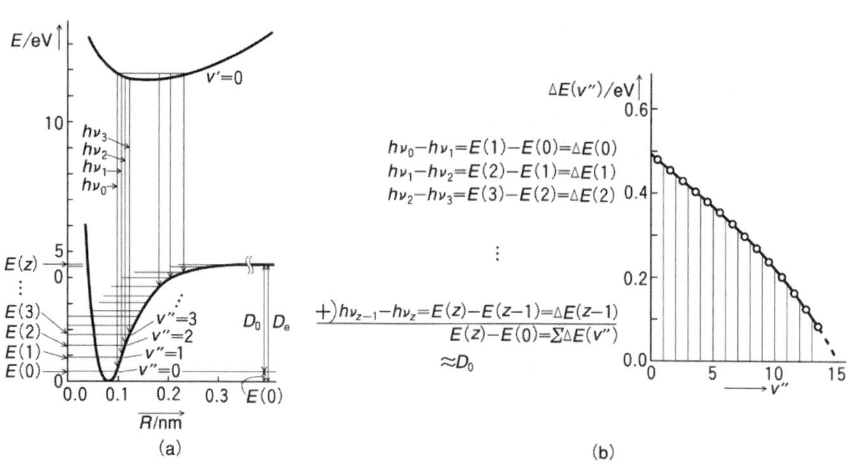

Fig. 7.1. (**a**) Schematic representation of an emission spectrum from an excited electronic state. (**b**) The plot of the spectral data to derive the dissociation energy

In Lyman's experiment, the H_2 molecule was excited by the bombardment of high energy electrons produced by electric discharge. When the molecules were excited to an electronic state lying about 12 eV above the ground electronic state, they emitted light with energies corresponding to the energy difference $h\nu_i$ between the ground vibrational state ($v' = 0$) in the excited electronic state and the various vibrationally excited states ($v'' = 0, 1, 2, \dots$) in the ground electronic state. According to Fig. 7.1(a), the difference between adjacent lines $h\nu_{i-1}$ and $h\nu_i$ is equal to the difference in the vibrational energies $E(i) - E(i-1)$. Thus, the plot of this difference against the vibrational quantum number v'' gives a stick diagram, as in Fig. 7.1(b), which is called the Birge-Sponer diagram. As is seen from this diagram, the sum of the stick heights is tantamount to the difference between the energy of the ground vibrational state in the ground electronic state and the energy of the dissociation limit as indicated by D_0 in Fig. 7.1(a). The sum of D_0 and $E(0)$ in Fig. 7.1(a) gives the dissociation energy D_e, where $E(0)$ is the difference between the potential minimum and the energy of $v'' = 0$, and is identified with the zero-point energy discussed in connection with Fig. 5.1.

E.E. Witmer (1926) whose work succeeded that of Lyman estimated the dissociation energy as $D_e = 4.35\,\mathrm{eV} \cong 420\,\mathrm{kJ\,mol^{-1}}$. This value is fairly close to the present value of $4.52\,\mathrm{eV} \cong 436\,\mathrm{kJ\,mol^{-1}}$. Figure 7.1(b) is actually drawn after the improved data of H. Beutler (1934), but the figure is basically similar to Witmer's in 1926.

No doubt, it was W. Heitler and F. London (1927) who discovered the essence of the chemical bonding in the H_2 molecule from the standpoint of quantum mechanics. However, it must be pointed out that they owe a debt to the experimental work accomplished by Witmer in the preceding year; Heitler and London themselves began their work in an attempt to estimate the van der Waals force between two separate H atoms by applying the newly born quantum mechanics. The concept of the van der Waals force had been put forth by J.D. van der Waals in the late nineteenth century before the advent of the new quantum theory. The force operates on all real atoms and molecules on the basis of classical electromagnetism. Its origin is the instantaneous fluctuation of the electric charge in atoms and molecules. The force is far-reaching and attractive so that the interaction energy due to this force acts to stabilize the assembly. Thus, it differentiates realistic atoms and molecules from the non-interacting particles constituting the virtual ideal gas. From various experiments and classical theories the van der Waals interaction energy had been estimated to be of the order of $10\,\mathrm{meV} \approx 1\,\mathrm{kJ\,mol^{-1}}$.

It is understandable that Heitler and London attempted to compute this interaction energy by quantum mechanics to confirm the validity of the new theory. The result of their computation, however, turned out to be orders of magnitude larger than the experimentally known van der Waals energy. The disappointment of the two men in their 20's, however, turned out to be rapture, thanks to the apt advice of A.J. Sommerfeld who, as a much more experienced scientist, suggested to them that their computational result actually might have contained in essence the binding energy of the H_2 molecule (A. Kojima, K. Higasi , 1975). Prior to this discovery, no one could explain the large binding energy of neutral diatomic molecules at all. However, the original computation by Heitler and London remained at giving a lower limit of the binding energy, which was about half of the dissociation energy of the H_2 molecule that had been obtained by Witmer experimentally as about 4.35 eV. Y. Sugiura (1927), being informed by Born with good timing, of the semiquantitative result of Heitler-London, quickly complemented their computation by carrying out an analytical computation for the interelectronic repulsion and got an improved energy of about 3 eV, which amounted to almost 2/3 of the experimental dissociation energy (cf. Sect. 7.2.1).

7.2 The Heitler-London Theory

The Heitler-London theory is also called the valence bond theory. It is one of the classics in molecular quantum mechanics, and its educative value is everlasting in studying the chemical bond between neutral atoms. In the following section, I will divide the explanation of the theory into two parts. In Sect. 7.2.1, the effects of electron spin are deliberately ignored; in Sect. 7.2.2, the discussion continues with the inclusion of the electron spin.

7.2.1 Eigenfunctions and Eigenvalues of Energy Ignoring the Effect of the Electron Spin

In the Born-Oppenheimer approximation the Hamiltonian and its eigenfunction of the H_2 molecule obey the following equation where the spatial coordinates of the two electrons are written as $\boldsymbol{r}_1$ and $\boldsymbol{r}_2$ and the distance between the two protons is regarded as a parameter.

$$\begin{aligned}
&\hat{H}(\boldsymbol{r}_1,\boldsymbol{r}_2,R)\Phi(\boldsymbol{r}_1,\boldsymbol{r}_2,R)\\
&=\left[\frac{-\hbar^2}{2m_\mathrm{e}}\Delta_1+\frac{-\hbar^2}{2m_\mathrm{e}}\Delta_2-\frac{e^2}{4\pi\varepsilon_0 r_{l1}}-\frac{e^2}{4\pi\varepsilon_0 r_{r1}}\right.\\
&\quad\left.-\frac{e^2}{4\pi\varepsilon_0 r_{l2}}-\frac{e^2}{4\pi\varepsilon_0 r_{r2}}+\frac{e^2}{4\pi\varepsilon_0 r_{12}}+\frac{e^2}{4\pi\varepsilon_0 R}\right]\Phi(\boldsymbol{r}_1,\boldsymbol{r}_2,R)\\
&=E\Phi(\boldsymbol{r}_1,\boldsymbol{r}_2,R)
\end{aligned}\tag{7.1}$$

The interactions between the two protons and two electrons are represented by the third through sixth terms in the brackets. The seventh and the eighth represent the repulsive interactions between the electrons and between the protons, respectively.

Now, consider that the two electrons are exchanged. The Hamiltonian in the brackets remains unchanged while the eigenfunction $\Phi(\boldsymbol{r}_1,\boldsymbol{r}_2,R)$ becomes $\Phi(\boldsymbol{r}_2,\boldsymbol{r}_1,R)$. Therefore, if $\Phi(\boldsymbol{r}_1,\boldsymbol{r}_2,R)$ is an eigenfunction of (7.1), then, $\Phi(\boldsymbol{r}_2, \boldsymbol{r}_1, R)$ is also an eigenfunction with the same eigenvalue of E. According to the statements in Sects. 4.1 and 6.1.1, the above relation means that $\Phi(\boldsymbol{r}_2, \boldsymbol{r}_1, R)$ must be at most a multiple of $\Phi(\boldsymbol{r}_1, \boldsymbol{r}_2, R)$, i.e., $\Phi(\boldsymbol{r}_2, \boldsymbol{r}_1, R) = \text{const} \times \Phi(\boldsymbol{r}_1, \boldsymbol{r}_2, R)$. Since a repetition of the exchange brings the system back to the original state, it is required that the constant must be ± 1. Thus, $\Phi(\boldsymbol{r}_2, \boldsymbol{r}_1, R)$ must be either $\Phi(\boldsymbol{r}_2, \boldsymbol{r}_1, R) = +\Phi(\boldsymbol{r}_1, \boldsymbol{r}_2, R)$ or $\Phi(\boldsymbol{r}_2, \boldsymbol{r}_1, R) = -\Phi(\boldsymbol{r}_1, \boldsymbol{r}_2, R)$.

We have to find eigenfunctions satisfying either one of the above two relations. As a trial, let us see whether the simplest function $\chi_l(1)\chi_r(2)$ (or, equivalently, $\chi_l(2)\chi_r(1)$) satisfies the above condition or not. However, the exchange induces the change $\chi_l(1)\chi_r(2) \rightarrow \chi_l(2)\chi_r(1)$ (or, equivalently, $\chi_l(2)\chi_r(1) \rightarrow \chi_l(1)\chi_r(2)$), so that the simple product which describes a state of each electron associated with one of the two protons cannot be a proper eigenfunction. However, it is obvious from the arguments reiterated so far that linear combinations of the simple products above, i.e., $\chi_l(1)\chi_r(2) \pm \chi_l(2)\chi_r(1)$, satisfy the requirement upon exchange; the plus combination leads to $\Phi(\boldsymbol{r}_2, \boldsymbol{r}_1, R) = +\Phi(\boldsymbol{r}_1, \boldsymbol{r}_2, R)$ and the minus combination to $\Phi(\boldsymbol{r}_2, \boldsymbol{r}_1, R) = -\Phi(\boldsymbol{r}_1, \boldsymbol{r}_2, R)$.

It is indeed these two functions that Heitler and London adopted as trial functions. Actually, they did not know that the proper eigenfunction must satisfy such a symmetry requirement upon the exchange of the two electrons; they knew only the necessity that the two H atoms should be treated equivalently and that the trial functions must be orthogonal as eigenfunctions for the eigenvalue equation (7.1). It was too early for them to construct their logic upon the now familiar concept of the indistinguishability of identical particles such as the electrons here and the symmetry requirement upon the exchange among them. Thus, it may be said that their choice of the above linear combinations was a fortuitous serendipity. After the trivial normalization, the functions that Heitler and London came up with are written as follows.

$$\Phi_{\mathrm{s,a}} = [\chi_l(1)\chi_r(2) \pm \chi_l(2)\chi_r(1)] \,/\, \sqrt{2(1 \pm S^2)} \tag{7.2}$$

The next step is, of course, to estimate the energy eigenvalues $E_{\mathrm{s,a}}$ corresponding to the above two trial functions. The energies are obtained by substituting (7.2) into (7.1) and by integrating with respect to the coordinates of electrons 1 and 2. To avoid the tediousness of dealing with similar terms we follow the technique after M. Karplus (1970). For this, the Hamiltonian in (7.1) is first rewritten as follows.

$$\hat{H} = \hat{H}_{\mathrm{I}} + \hat{V}_{\mathrm{I}} = \hat{H}_{\mathrm{II}} + \hat{V}_{\mathrm{II}} \tag{7.3}$$

$$\hat{H}_{\mathrm{I}} = \left(\frac{-\hbar^2}{2m_{\mathrm{e}}} \Delta_1 - \frac{e^2}{4\pi\varepsilon_0 r_{l1}} \right) + \left(\frac{-\hbar^2}{2m_{\mathrm{e}}} \Delta_2 - \frac{e^2}{4\pi\varepsilon_0 r_{r2}} \right) \tag{7.4}$$

$$\hat{V}_{\mathrm{I}} = \frac{e^2}{4\pi\varepsilon_0} \left(-\frac{1}{r_{r1}} - \frac{1}{r_{l2}} + \frac{1}{r_{12}} + \frac{1}{R} \right) \tag{7.5}$$

$$\hat{H}_{\mathrm{II}} = \left(\frac{-\hbar^2}{2m_{\mathrm{e}}}\Delta_1 - \frac{e^2}{4\pi\varepsilon_0 r_{r1}}\right) + \left(\frac{-\hbar^2}{2m_{\mathrm{e}}}\Delta_2 - \frac{e^2}{4\pi\varepsilon_0 r_{l2}}\right) \tag{7.6}$$

$$\hat{V}_{\mathrm{II}} = \frac{e^2}{4\pi\varepsilon_0}\left(-\frac{1}{r_{l1}} - \frac{1}{r_{r2}} + \frac{1}{r_{12}} + \frac{1}{R}\right) \tag{7.7}$$

The next step is to obtain $E_{\mathrm{s,a}}$, therefore the following expression is prepared first.

$$\hat{H}\Phi_{\mathrm{s,a}} = \hat{H}\left[\chi_l(1)\chi_r(2) \pm \chi_l(2)\chi_r(1)\right]/\sqrt{2(1 \pm S^2)} \tag{7.8}$$

Then, as a little trick, the Hamiltonian in the right-hand side is rewritten as follows in reference to (7.3).

$$\hat{H}\Phi_{\mathrm{s,a}} = \left[(\hat{H}_{\mathrm{I}} + \hat{V}_{\mathrm{I}})\chi_l(1)\chi_r(2) \pm (\hat{H}_{\mathrm{II}} + \hat{V}_{\mathrm{II}})\chi_l(2)\chi_r(1)\right]/\sqrt{2(1 \pm S^2)} \tag{7.9}$$

Accordingly, the expression sought, $E_{\mathrm{s,a}}$, can be expressed as below.

$$\begin{aligned} E_{\mathrm{s,a}} = & \iint \{\chi_l(1)\chi_r(2) \pm \chi_l(2)\chi_r(1)\} \\ & \times(\hat{H}_{\mathrm{I}} + \hat{V}_{\mathrm{I}})\chi_l(1)\chi_r(2)\mathrm{d}v_1\mathrm{d}v_2/[2(1 \pm S^2)] \\ \pm & \iint \{\chi_l(1)\chi_r(2) \pm \chi_l(2)\chi_r(1)\} \\ & \times(\hat{H}_{\mathrm{II}} + \hat{V}_{\mathrm{II}})\chi_l(2)\chi_r(1)\mathrm{d}v_1\mathrm{d}v_2/[2(1 \pm S^2)] \end{aligned} \tag{7.10}$$

The expansion of the above equation produces a number of terms, but the overall result can be simplified because the exchange of electrons 1 and 2 forces the exchange of $\hat{V}_{\mathrm{I}}$ and $\hat{V}_{\mathrm{II}}$ and because the terms involving both $\hat{H}_{\mathrm{I}}$ and $\hat{H}_{\mathrm{II}}$ simply end up giving the eigenvalue of the 1s orbital of the H atom, i.e., $E_{1\mathrm{s}}$. This simplified result is:

$$\begin{aligned} E_{\mathrm{s,a}} = 2E_{1\mathrm{s}} & \pm \iint \chi_l(1)\chi_r(2)\hat{V}_{\mathrm{I}}\chi_l(1)\chi_r(2)\mathrm{d}v_1\mathrm{d}v_2/(1 \pm S^2) \\ & \pm \iint \chi_l(2)\chi_r(1)\hat{V}_{\mathrm{I}}\chi_l(1)\chi_r(2)\mathrm{d}v_1\mathrm{d}v_2/(1 \pm S^2) \end{aligned} \tag{7.11}$$

The second and third integrals of the above equation are called the Coulomb integral and the exchange integral in the Heitler-London theory and are often denoted by the symbols J and K, respectively (unfortunately, some solid state physicists use the symbols Q and J, which correspond to the above J and K. The readers are warned to be careful about the same symbol J which might be confusing depending on authors). In this book the two symbols are used for the meanings as below.

$$\begin{aligned} J &= \iint \chi_l(1)\chi_r(2)\hat{V}_{\mathrm{I}}\chi_l(1)\chi_r(2)\mathrm{d}v_1\mathrm{d}v_2 \\ K &= \iint \chi_l(2)\chi_r(1)\hat{V}_{\mathrm{I}}\chi_l(1)\chi_r(2)\mathrm{d}v_1\mathrm{d}v_2 \end{aligned} \tag{7.12}$$

According to the notations above, the energies in (7.11) can be concisely expressed as:

$$E_{\rm s} = 2E_{1\rm s} + \frac{J+K}{1+S^2}, \quad E_{\rm a} = 2E_{1\rm s} + \frac{J-K}{1-S^2} \tag{7.13}$$

The above results for the H_2 molecule correspond to the results for the H_2^+ system in (6.24) reproduced below for comparison.

$$E_{\rm s} = E_{1\rm s} + \frac{J+K}{1+S}, \quad E_{\rm a} = E_{1\rm s} + \frac{J-K}{1-S} \tag{6.24}$$

It must be noted, however, that the meanings of J and K in (7.13) are different from those in (6.24) as mentioned in connection with (6.25) and (6.26).

In the computations of J, K, and S in (7.12) and (7.13), Heitler and London employed the 1s orbital of the H atom for χ_l and χ_r. Such a constriction is unnatural because the H atom in the H_2 molecule cannot be the same as in the isolated H atom. However, their approach is understandable because the initial motivation of their study was to estimate the van der Waals interaction energy with an implicit presumption that the two atoms are far enough apart to be approximated as two individual atoms.

As to the actual computation of $E_{\rm s,a}$, even under the forced assumption of $\chi_l = \chi_r =$ (1s orbital of the H atom) as in the simpler H_2^+ system, the energy expression becomes more complicated than that in (6.27). This is because in the H_2 molecule new terms, $\iint \chi_l(1)\chi_r(2)(1/r_{12})\chi_l(1)\chi_r(2)\mathrm{d}v_1\mathrm{d}v_2$ and $\iint \chi_l(2)\chi_r(1)(1/r_{12})\chi_l(1)\chi_r(2)\mathrm{d}v_1\mathrm{d}v_2$ representing the interelectronic repulsive interaction, are added. These new terms cannot be obtained by elementary integrals to lead to a result such as in (6.27). As a matter of fact, Heitler and London refrained from evaluating these integrals and only gave a lower limit of the absolute values of $E_{\rm s,a}$. As mentioned in Sect. 7.1, it was Sugiura who performed the above integrals within the same year as that of the work by Heitler and London (1927).

As is seen from Fig. 7.2(a), the general features of the computational result of Heitler-London-Sugiura turned out to be similar to the features of the H_2^+ system shown in Fig. 6.2(a). Thus, in the H_2 molecule also, the functions $\Phi_{\rm s}$ and $\Phi_{\rm a}$ in (7.2) correspond to the bonding and the anti-bonding states, respectively, as in the case of the H_2^+ system.

The discrepancy between $E_{\rm s,a}^{\rm HLS}$ and $E_{\rm s,a}^{\rm KW}$ is mainly due to the unnatural restriction of χ_l and χ_r to the 1s orbital of the isolated H atom. Nevertheless, the energy of $E_{\rm s}^{\rm HLS}$ amounts to almost two-thirds of the experimental energy. This indicates that the original consideration of Heitler and London that the two H atoms should participate equally in the bonding cuts to the essence of the bonding in the H_2 molecule.

To be more concrete, the essential stability of the bonding is mainly gained by the contribution of the exchange integral which is derived from the superposition of the two degenerate states $\chi_l(1)\chi_r(2)$ and $\chi_l(2)\chi_r(1)$, with regard

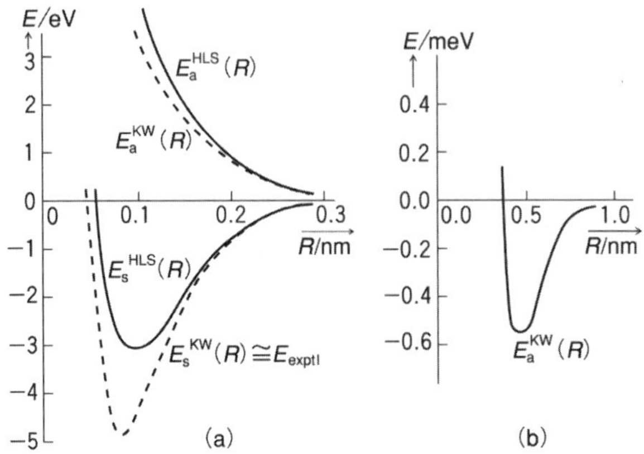

Fig. 7.2. (**a**) The solid curves represent the result of the computation by Heitler-London-Sugiura for the ground electronic (E_s) and the first excited electronic state (E_a). The broken curves are the results of later sophisticated computations by Kolos and Wolniewicz for the two states. Their results are in an almost complete agreement with the result (E_{exptl}) of the elaborate experiments by Herzberg and others (cf. Sect. 7.3). (**b**) The explanation for the E_a^{KW} curve at large internuclear distances will be made in Sect. 7.3

to the exchange of electrons 1 and 2. It is of the utmost importance to fully understand the importance of the superposition of state in quantum mechanics as is shown in this section for the H_2 system and in Chap. 6 for the H_2^+ ion.

Problem

7.1 Derive (7.11) from (7.9).

7.2.2 Eigenfunctions and Eigenvalues Inclusive of the Electron Spin

In the preceding section the electron spin was ignored in describing the H_2 molecule to avoid the added complexity at the beginning level. For complete understanding the electron spin must be included. Thus, this section will focus on the role of the electron spin.

Since the two electrons in the H_2 molecule are fermions (cf. Sect. 4.1), they must be subjected to the Pauli exclusion principle. The problem of this section is how to take into account this universal requirement in the present system of the H_2 molecule.

As discussed in Sect. 4.1, the function $\Phi(\boldsymbol{r}_1, \boldsymbol{r}_2, R)$ to describe multi-electron systems is supposed to be expressible in terms of the spin orbitals $\phi(\boldsymbol{r}\sigma)$ of the individual electrons. However, since the spatial coordinate $\boldsymbol{r}$ and the spin coordinate σ belong to two completely different spaces, the Hamiltonian $\hat{H}$ and the two electron spin operators $\hat{\boldsymbol{s}}^2$ and $\hat{s}_z$ must be commutative, i.e., we have the relations $[\hat{H}, \hat{\boldsymbol{s}}^2] = 0$ and $[\hat{H}, \hat{s}_z] = 0$ (cf. Sects. 3.4.4 and 2.1.12). This commutativity means that the functions Φ_s and Φ_a obtained so far as the eigenfunctions of $\hat{H}$ must also be the simultaneous eigenfunctions of $\hat{\boldsymbol{s}}^2$ and $\hat{s}_z$. Thus, the states described by the eigenfunctions $\Phi(\boldsymbol{r}_1\sigma_1, \boldsymbol{r}_2\sigma_2, R)$ should be able to be classified not only by the energy but also by the eigenvalue of the electron spin operators.

In order to confirm what is stated above for the real world system of the H_2 molecule, the eigenvalue problem for the electron spin operators will be examined to reach the final results as shown in (7.22) and (7.23).

In one-electron systems the eigenvalues of the operators $\hat{\boldsymbol{s}}^2$ and $\hat{s}_z$ are $3\hbar^2/4$ and $\pm\hbar/2$, respectively, as given in (3.36) and (3.37). In the H_2 molecule, the spin state of the two electrons must be determined. Each of the two electrons can be either in the spin-up state α or in the spin-down state β as stated in Sect. 3.4.3. The two states are energetically degenerate in the absence of external magnetic fields. Therefore, the electron spin component of the total eigenfunction $\Phi(\boldsymbol{r}_1\sigma_1, \boldsymbol{r}_2\sigma_2, R)$ is to be described by any one of the following four combinations $\alpha(1)\alpha(2)$, $\alpha(1)\beta(2)$, $\beta(1)\alpha(2)$, and $\beta(1)\beta(2)$.

Let us first examine the effect of the z-axis component of the spin operator written as $\hat{S}_z$ upon the above four combinations. The operator $\hat{S}_z$ is a simple sum of $\hat{s}_{z1}$ and $\hat{s}_{z2}$ because electrons 1 and 2 are independent. As an example, we get the result below, which shows that the eigenvalue of $\hat{S}_z$ for the first member of the combinations is $+\hbar$.

$$\begin{aligned}\hat{S}_z\alpha(1)\alpha(2) &= (\hat{s}_{z1} + \hat{s}_{z2})\alpha(1)\alpha(2) = \alpha(2)\hat{s}_{z1}\alpha(1) + \alpha(1)\hat{s}_{z2}\alpha(2) \\ &= +\frac{\hbar}{2} \times 2\alpha(1)\alpha(2) = +\hbar\alpha(1)\alpha(2) \qquad (7.14)\end{aligned}$$

Similarly, we get $-\hbar$ as the eigenvalue for the member of $\beta(1)\beta(2)$. As to the remaining two members $\alpha(1)\beta(2)$ and $\beta(1)\alpha(2)$, the corresponding eigenvalues written as $m_{s1} + m_{s2} \equiv M_S$ become 0, i.e., the two members are doubly degenerate. Therefore, the simultaneous eigenfunctions for the commutative operators, $\hat{S}_z$ and $\hat{\boldsymbol{S}}^2 \equiv (\hat{\boldsymbol{s}}_1 + \hat{\boldsymbol{s}}_2)^2$, must be a linear combination of $\alpha(1)\beta(2)$ and $\beta(1)\alpha(2)$ according to the general argument in Sect. 2.1.7. Putting this linear combination into the form as shown in (7.15) we attempt to determine the eigenvalue K of $\hat{\boldsymbol{S}}^2$ in (7.16) and the coefficients c_I and c_II.

$$\Phi(M_S = 0) = c_\mathrm{I}\alpha(1)\beta(2) + c_\mathrm{II}\alpha(2)\beta(1) \equiv c_\mathrm{I}\phi_\mathrm{I} + c_\mathrm{II}\phi_\mathrm{II} \qquad (7.15)$$

$$\hat{\boldsymbol{S}}^2\Phi(M_S = 0) = K\Phi(M_S = 0) \qquad (7.16)$$

For this purpose we first substitute $\Phi(M_S = 0) = c_\mathrm{I}\phi_\mathrm{I} + c_\mathrm{II}\phi_\mathrm{II}$ in (7.16), and then, multiply both sides of (7.16) from the left by ϕ_I^* and by ϕ_II^* and integrate with respect to the spin coordinate to obtain K, c_I, and c_II, as usual.

The integrals are as follows: $\iint \phi_{\mathrm{I}}^* \hat{\boldsymbol{S}}^2 \phi_{\mathrm{I}} \mathrm{d}\sigma_1 \mathrm{d}\sigma_2 \equiv S_{\mathrm{I,I}}^2$, $\iint \phi_{\mathrm{I}}^* \hat{\boldsymbol{S}}^2 \phi_{\mathrm{II}} \mathrm{d}\sigma_1 \mathrm{d}\sigma_2 \equiv S_{\mathrm{I,II}}^2$, and $\iint \phi_{\mathrm{II}}^* \hat{\boldsymbol{S}}^2 \phi_{\mathrm{I}} \mathrm{d}\sigma_1 \mathrm{d}\sigma_2 \equiv S_{\mathrm{II,I}}^2$, $\iint \phi_{\mathrm{II}}^* \hat{\boldsymbol{S}}^2 \phi_{\mathrm{II}} \mathrm{d}\sigma_1 \mathrm{d}\sigma_2 \equiv S_{\mathrm{II,II}}^2$ and the secular equation to be solved is in (7.17).

$$\begin{vmatrix} S_{\mathrm{I,I}}^2 - K & S_{\mathrm{I,II}}^2 \\ S_{\mathrm{II,I}}^2 & S_{\mathrm{II,II}}^2 - K \end{vmatrix} = 0 \tag{7.17}$$

The four integrals above are obtained as follows: since the spin operator $\hat{\boldsymbol{s}}$ of a single electron is a vectorial operator with three components $\hat{s}_x$, $\hat{s}_y$, and $\hat{s}_z$, the expansion of $\hat{\boldsymbol{S}}^2$ can be written as in (7.18) where the dot signifies the dot product (inner product) of the two different vectorial operators $\hat{\boldsymbol{s}}_1$ and $\hat{\boldsymbol{s}}_2$.

$$\begin{aligned} \hat{\boldsymbol{S}}^2 &= (\hat{\boldsymbol{s}}_1 + \hat{\boldsymbol{s}}_2)^2 = \hat{\boldsymbol{s}}_1^2 + \hat{\boldsymbol{s}}_2^2 + 2\hat{\boldsymbol{s}}_1 \cdot \hat{\boldsymbol{s}}_2 \\ &= \hat{\boldsymbol{s}}_1^2 + \hat{\boldsymbol{s}}_2^2 + 2(\hat{s}_{1x}\hat{s}_{2x} + \hat{s}_{1y}\hat{s}_{2y} + \hat{s}_{1z}\hat{s}_{2z}) \end{aligned} \tag{7.18}$$

In reference to (7.18) and the properties of the eigenfunctions α and β given in Sects. 3.4.3 and 3.4.4 (in particular, (3.39)), the first element in (7.17), $S_{\mathrm{I,I}}^2$, is calculated as in (7.19) below.

$$\begin{aligned} &\iint \alpha^*(1)\beta^*(2)\big[\hat{\boldsymbol{s}}_1^2\alpha(1)\beta(2) + \alpha(1)\hat{\boldsymbol{s}}_2^2\beta(2) + 2\hat{s}_{1x}\alpha(1)\hat{s}_{2x}\beta(2) \\ &\quad +2\hat{s}_{1y}\alpha(1)\hat{s}_{2y}\beta(2) + 2\hat{s}_{1z}\alpha(1)\hat{s}_{2z}\beta(2)\big]\mathrm{d}\sigma_1\mathrm{d}\sigma_2 \\ &= \iint \alpha^*(1)\beta^*(2)\big[(3\hbar^2/4)\alpha(1)\beta(2) + (3\hbar^2/4)\alpha(1)\beta(2) \\ &\quad +2(+\hbar/2)\beta(1)(+\hbar/2)\alpha(2) + 2(+\mathrm{i}\hbar/2)\beta(1)(-\mathrm{i}\hbar/2)\alpha(2) \\ &\quad +2(+\hbar/2)\alpha(1)(-\hbar/2)\beta(2)\big]\mathrm{d}\sigma_1\mathrm{d}\sigma_2 \\ &= 3\hbar^2/4 + 3\hbar^2/4 - 2\hbar^2/4 = \hbar^2 \end{aligned} \tag{7.19}$$

From similar calculations all the remaining elements $S_{\mathrm{I,II}}^2$, $S_{\mathrm{II,I}}^2$, and $S_{\mathrm{II,II}}^2$, are also found to be equal to $\hbar^2$, so that (7.17) is explicitly given as below.

$$\begin{vmatrix} \hbar^2 - K & \hbar^2 \\ \hbar^2 & \hbar^2 - K \end{vmatrix} = 0 \tag{7.20}$$

As a result, the eigenvalue K is determined to be either $2\hbar^2$ or 0. The former leads to the relation $c_{\mathrm{I}} = c_{\mathrm{II}}$, while the latter gives $c_{\mathrm{I}} = -c_{\mathrm{II}}$. Thus, the sought function $\Phi(M_S = 0)$ in (7.15) is obtained as follows after normalization:

$$\Phi_{\mathrm{s,a}}(M_S = 0) = (1/\sqrt{2})[\alpha(1)\beta(2) \pm \alpha(2)\beta(1)] \tag{7.21}$$

Since the eigenvalue K and the spin quantum number S of the H_2 molecule are related through $K = S(S+1)\hbar^2$, the two eigenvalues $K = 2\hbar^2$ and 0 correspond, respectively, to $S = 1$ and $S = 0$ and the spin multiplicity $2S+1$ is found to be a triplet and singlet, respectively.

Thus, we can conclude that the electron spin part of the eigenfunction of the H_2 molecule includes the components $\alpha(1)\alpha(2)$ ($M_S = +1$), $\beta(1)\beta(2)$ ($M_S = -1$), and $(1/\sqrt{2})[\alpha(1)\beta(2) \pm \alpha(2)\beta(1)]$ ($M_S = 0$).

The products of these components with the spatial components of the eigenfunctions in (7.2) give the overall eigenfunctions which must satisfy the requirement of the Pauli exclusion principle. It is apparent that the following four products in (7.22) and (7.23) are the correct functions where the superscripts 1 and 3 indicate that the first of the four is for the singlet state and the remaining three are for the triplet.

$$
\begin{aligned}
{}^1\Phi(\boldsymbol{r}_1\sigma_1, \boldsymbol{r}_2\sigma_2) &= \frac{1}{\sqrt{2(1+S^2)}}[\chi_l(1)\chi_r(2) + \chi_l(2)\chi_r(1)] \\
&\quad \times \frac{1}{\sqrt{2}}[\alpha(1)\beta(2) - \alpha(2)\beta(1)]
\end{aligned} \tag{7.22}
$$

$$
\begin{aligned}
{}^3\Phi(\boldsymbol{r}_1\sigma_1, \boldsymbol{r}_2\sigma_2) &= \frac{1}{\sqrt{2(1-S^2)}}[\chi_l(1)\chi_r(2) - \chi_l(2)\chi_r(1)] \\
&\quad \times \begin{cases} \alpha(1)\alpha(2) \\ \frac{1}{\sqrt{2}}[\alpha(1)\beta(2) + \alpha(2)\beta(1)] \\ \beta(1)\beta(2) \end{cases}
\end{aligned} \tag{7.23}
$$

In (7.22) the spatial component is symmetric and the spin component is anti-symmetric against the exchange of electrons 1 and 2, whereas the reverse is true in (7.23).

In Sect. 4.1 it was shown that the Pauli exclusion principle is automatically and concisely taken into account if the eigenfunction is expressed in the form of a Slater determinant. Therefore, (7.22) and (7.23) should be expressible in terms of a determinant. In fact, rearrangements in (7.22) and (7.23) lead to the following representations with the use of the notation in (4.5).

$$
{}^1\Phi(\boldsymbol{r}_1\sigma_1, \boldsymbol{r}_2\sigma_2) = \frac{1}{\sqrt{2(1+S^2)}}\left[\|\chi_l(1)\overline{\chi}_r(2)\| - \|\overline{\chi}_l(1)\chi_r(2)\|\right] \tag{7.24}
$$

$$
{}^3\Phi(\boldsymbol{r}_1\sigma_1, \boldsymbol{r}_2\sigma_2) = \frac{1}{\sqrt{2(1-S^2)}} \times \begin{cases} \|\chi_l(1)\chi_r(2)\| \\ \|\chi_l(1)\overline{\chi}_r(2)\| + \|\overline{\chi}_l(1)\chi_r(2)\| \\ \|\overline{\chi}_l(1)\overline{\chi}_r(2)\| \end{cases} \tag{7.25}
$$

In Sect. 7.2.1, the electron spin component was ignored and it was concluded that the symmetric and the anti-symmetric spatial functions correspond respectively to the bonding and the anti-bonding states. Inspection of (7.22) through (7.25) reveals that such a classification of the states can also be made by referring to the difference in the spin components.

That the electron spin component can differentiate the two states led Heitler and London to state in their original paper that "and as it were the Pauli exclusion principle as a watch forbids by large repulsion hydrogen atoms to approach each other".

In Sect. 1.3 it was stated that the energy associated with the electron spin in H_2 is trivially small. As the above example shows, however, the electron spin plays, though indirectly, a dictating role in the formation of the chemical bond.

7.3 Development of the Heitler-London Theory

As mentioned in Sect. 7.1, the motivation for Heitler and London was, at first, to estimate the van der Waals interaction energy between two H atoms which were assumed to be suitably far apart. Therefore, it is not unreasonable that they adopted the 1s orbitals $\chi_l(1)$ and $\chi_r(2)$ of the H atoms in order to approximate the two-atom system by the functions $\Phi_{\mathrm{s,a}}$ in (7.2). When the two atoms are close enough to be regarded as a molecular complex, the form of (7.2) may be interpreted as representing the fact that each atom supplies its single electron to be shared equally by the two protons. In that sense (7.2) is associated with the name of the covalent bond.

However, it is unnatural to assume that the system can be approximated by the 1s atomic orbitals even when the two atoms approach to merge in a H_2 molecule. At short distances, the electron assumed to retain its 1s orbital functional form would be affected by the electric field of the counterpart proton, which means there would be an increase in the effective nuclear charge from $+Z(=1)e$ to $+Z(>1)e$. It is also imaginable that occasionally the two electrons may be localized on one of the two protons to yield the ionic bond situation, which can be denoted as $H_l^- H_r^+$ and $H_l^+ H_r^-$, or $\chi_l(1)\chi_l(2)$ and $\chi_r(1)\chi_r(2)$ in conformity with the expression in (7.2). Because of such limitations, the original Heiler-London theory has been subjected to a number of improvements such as the addition of non-spherical atomic orbitals to increase the flexibility of the basis functions.

A resolute departure from the atomic orbital approximation was tried by H.M. James and A.P. Coolidge (1933) which was succeeded to W. Kolos and L. Wolniewicz (1965) among others: now that the symmetry consideration is of utmost import in homodiatomic systems like H_2^+ and H_2, these people employed functions of the form $[\varphi_i(1,2) \pm \varphi_i(2,1)]$ as the basis functions where the two-electron function φ_i bears no relation to atomic orbitals but contains the interelectronic distance r_{12} as an explicit variable in addition to the coordinates of individual electrons. Furthermore, the overall function was constructed by a linear combination of the above symmetry-adapted functions as $\sum c_i[\varphi_i(1,2) \pm \varphi_i(2,1)]$.

Based on these fundamentally different and flexible trial functions Kolos and Wolniewicz performed thoroughgoing computations within the framework of the calculus of variation. Since they took into account part of the kinetic energy of the protons and the effects of the electron spin and other relativistic terms as perturbers in addition to the major Hamiltonian in (7.1), their computational results are beyond the limits of the Born-Oppenheimer approximation. As a result, their predicted analyses reached such a high level of approximation as to be indistinguishable from the authentic experimental results by Herzberg (1970) and I. Dabrowski (1984) (cf. Fig. 7.2(a)). This book is not intended to delve into the details of these sophisticated studies. However, considering that the original motivation of Heitler and London was to quantify the van der Waals interaction between two atoms, part of the work by Kolos and Wolniewicz relevant to this topic is briefly excerpted below.

The van der Waals interaction is universally operative among all atoms and molecules. However, the magnitude of the interaction is much smaller than the exchange interaction between two atoms, whose electron clouds overlap significantly. Analysis of the results of Kolos and Wolniewicz (1974) reveals that the van der Waals interaction contributes to stabilize the H_2 molecule in both the bonding *and* the anti-bonding states. In the former the small stabilization is overshadowed by the overwhelmingly large stabilization due to the exchange interaction, but in the latter triplet state the small stabilization due to the van der Waals interaction is manifested as a very shallow minimum in the repulsive curve in Fig. 7.2(b). It is to be noted that in Fig. 7.2(b) the scale of the ordinate is roughly four orders of magnitude smaller than that of Fig. 7.2(a) and that the minimum occurs at the large internuclear distance of about 0.413 nm with an energy of about 0.557 meV relative to the dissociation limit.

The above stabilization energy for the triplet state is even smaller than the thermal energy at room temperatures, $kT \approx 25\,\text{meV}$ where k stands for the Boltzmann constant. Therefore, the contribution of the van der Waals stabilization is practically ignorable for the H_2 molecule in the 'repulsive' triplet state. However, the universality of the van der Waals interaction at distances much larger than the distance of chemical bonds has a potential significance in, e.g., the formation of various molecular clusters. The interaction may also play a delicate but important role in various phenomena in biological systems, for example. In this sense the original intention of Heitler and London to investigate the van der Waals interaction should be regarded as right and appropriate even today.

The detection of such a delicate stabilization at a long internuclear distance as shown in Fig. 7.2(b) is exceptional thanks to the availability of symmetry-adapted flexible functions of the form $\sum c_i[\varphi_i(1,2) \pm \varphi_i(2,1)]$ in confocal elliptic coordinates. Except for the special cases of H_2^+ and H_2, almost all molecules are analyzed in terms of molecular orbitals approximated by linear combination of atomic orbitals, or more generally, of basis functions in rectangular coordinates and by applying the criterion of the energy minimum in Ritz's calculus of variation. The best solution obtained by this technique, however, does not necessarily guarantee the adequacy of the trial function at distances where any subtle interaction is manifested. This general problem of the LCAO-MO approximation remains to be resolved (cf. Chap. 8).

7.4 Molecular Orbital Treatments of the Hydrogen Molecule

Since the H_2^+ ion is a single-electron system, the LCAO orbital itself becomes an approximate eigenfunction of the system. In contrast, in the case of the two-electron system of H_2, one must first consider the Pauli exclusion principle before finding any approximate eigenfunction. In Sect. 7.2.2 one method of obtaining such anti-symmetrized total eigenfunctions as in (7.22) through (7.25) was demonstrated.

However, in the approximate eigenfunctions of Sect. 7.2.2, there is an intrinsic limitation in improving the degree of approximation because the functions are fixed as AO-based Heitler-London type. In the other extreme case, i.e., in the case of the Kolos-Wolniewicz approach, while the accuracy of the function could be improved with an increase in the number of parameters in the calculus of variation, this method cannot be applied easily to molecules other than the simplest homodiatomic molecule H_2. Although the work of Kolos and Wolniewicz on the H_2 molecule is an important milestone in molecular quantum mechanics, it is too rigorous for most problems involving other molecules.

As a compromise, we will examine the LCAO-MO approximation in the analysis of the two-electron H_2 system. Although the primitive LCAO-MO approximation is inflexible as stated above, substitution of the native AO with some prepared basis functions facilitates an increase in the accuracy of the approximation by improving on the basis functions within the framework of Ritz's calculus of variation. Importantly, this approach can be applied not merely to diatomic but to any polyatomic molecules (cf. Chap. 8).

The above stated substitution of the atomic orbitals with proper basis functions is one of the central issues in textbooks and manuals concerned with quantum chemical computations. This book is not intended to pursue such technical methodology but to introduce to the readers the versatility of the LCAO-MO approximation. The simplest H_2 system is taken as an example in this chapter. Our starting point is the same as (6.6) prepared for the description of the one-electron system in Sect. 6.3.

$$\begin{aligned}\varphi_{\mathrm{s}} &= [\chi_l(\boldsymbol{r},\boldsymbol{R}) + \chi_r(\boldsymbol{r},\boldsymbol{R})]\,/\sqrt{2(1+S)}\\ \varphi_{\mathrm{a}} &= [\chi_l(\boldsymbol{r},\boldsymbol{R}) - \chi_r(\boldsymbol{r},\boldsymbol{R})]\,/\sqrt{2(1-S)}\end{aligned} \tag{7.26}$$

In the above, $\chi_l(\boldsymbol{r},\boldsymbol{R}) \equiv \chi_l(\boldsymbol{r}+\boldsymbol{R}/2)$ and $\chi_r(\boldsymbol{r},\boldsymbol{R}) \equiv \chi_r(\boldsymbol{r}-\boldsymbol{R}/2)$ represent the 1s orbitals $1\mathrm{s}(|\boldsymbol{r}+\boldsymbol{R}/2|)$ and $1\mathrm{s}(|\boldsymbol{r}-\boldsymbol{R}/2|)$ for the H atom in its ground electronic state (cf. Fig. 6.1). The right-hand side of (7.26) is the same as the right-hand side of (6.6). In (6.6) the one-electron orbital itself was regarded as an approximate total eigenfunction of the system so that the capital letters Φ_{s} and Φ_{a} were used. In contrast, in (7.26) above, each of the two functions is regarded as an approximate one-electron trial orbital so that the small letters are used instead.

We now consider the Pauli exclusion principle. It is natural to conceive that in the ground electronic state of the molecule the two electrons prefer the orbital of lower energy with their electron spins being anti-parallel in deference to the principle. Therefore, the approximate eigenfunction for the ground electronic state $\Phi_{\mathrm{G}}(1,2)$ is expressed in terms of the Slater determinant as follows.

$$\begin{aligned}\Phi_{\mathrm{G}}(1,2) = \Phi_{\mathrm{G}}(\boldsymbol{r}_1\sigma_1, \boldsymbol{r}_2\sigma_2) &= \frac{1}{\sqrt{2!}}\begin{vmatrix}\varphi_{\mathrm{s}}(\boldsymbol{r}_1)\alpha(\sigma_1) & \varphi_{\mathrm{s}}(\boldsymbol{r}_1)\beta(\sigma_1)\\ \varphi_{\mathrm{s}}(\boldsymbol{r}_2)\alpha(\sigma_2) & \varphi_{\mathrm{s}}(\boldsymbol{r}_2)\beta(\sigma_2)\end{vmatrix}\\ &= \|\varphi_{\mathrm{s}}(\boldsymbol{r}_1)\alpha(\sigma_1)\varphi_{\mathrm{s}}(\boldsymbol{r}_2)\beta(\sigma_2)\| = \|\varphi_{\mathrm{s}}\overline{\varphi}_{\mathrm{s}}\|\end{aligned} \tag{7.27}$$

Expansion of (7.27) leads to (7.28) below.

$$\Phi_{\mathrm{G}}(1,2) = \varphi_{\mathrm{s}}(\boldsymbol{r}_1)\varphi_{\mathrm{s}}(\boldsymbol{r}_2)[\alpha(\sigma_1)\beta(\sigma_2) - \alpha(\sigma_2)\beta(\sigma_1)]/\sqrt{2} \tag{7.28}$$

Since (7.28) changes sign with the exchange of electrons 1 and 2, it is confirmed that the Pauli exclusion principle is satisfied. Furthermore, according to the statement in Sect. 7.2.2, the function $\Phi_{\mathrm{G}}(1,2)$ is known to express the singlet state (cf. (7.22)).

Substituting (7.26) into (7.28) the same function can be expressed also as in the following.

$$\begin{aligned}\Phi_{\mathrm{G}}(1,2) = &\frac{1}{2\sqrt{2}(1+S)}[\chi_l(1)\chi_l(2) + \chi_l(1)\chi_r(2) + \chi_l(2)\chi_r(1) + \chi_r(1)\chi_r(2)]\\ &\times[\alpha(\sigma_1)\beta(\sigma_2) - \alpha(\sigma_2)\beta(\sigma_1)]\end{aligned} \tag{7.29}$$

The above equation corresponds to (7.22) which describes the ground electronic state of the molecule in terms of the Heitler-London approximation. Comparison of (7.22) and (7.29) reveals that only the second and the third terms in the brackets of (7.29) appear in (7.22). These two terms can be interpreted to represent the state in which electrons 1 and 2 are shared equally by the two protons. Such a state may be identified with the covalent bond mentioned in Sect. 7.3. As for the first and fourth terms in the brackets of (7.29), they can be assigned to the ionic bond in Sect. 7.3. It is apparent that the ground electronic state of the actual H_2 molecule cannot be approximated simply by the covalent bond as in (7.22), because the two electrons may occasionally be localized instantly on one of the protons to cause the ionic bond. On the other hand, the equal contribution of the ionic bond to that of the covalent bond implied by the four terms in (7.29) cannot be regarded as acceptable in view of the various properties of the actual H_2 molecule. After all, neither the Heitler-London nor the LCAO-MO approximation can be considered as a satisfactory approximation when taken by themselves individually.

Anyway, the expectation value associated with the trial function in (7.29) for the ground electronic state can be calculated in a manner similar to the procedure used for the trial function of the Heitler-London type in Sect. 7.2.1. As for the excited triplet state Φ_{E}, it is natural to assume that one of the two electrons remains in φ_s of the lower energy, while the other is promoted to φ_a. Referring to (7.28) for the ground singlet state and to (7.23) for the electron spin component, the trial function for the excited triplet state is now written as follows.

$$\Phi_E(1,2) = \{\varphi_s(\boldsymbol{r}_1)\varphi_a(\boldsymbol{r}_2) - \varphi_a(\boldsymbol{r}_1)\varphi_s(\boldsymbol{r}_2)\} \times \begin{cases} \alpha(\sigma_1)\alpha(\sigma_2) \\ [\alpha(\sigma_1)\beta(\sigma_2) + \alpha(\sigma_2)\beta(\sigma_1)]/\sqrt{2} \\ \beta(\sigma_1)\beta(\sigma_2) \end{cases} \tag{7.30}$$

The actual computation of the expectation energies for $\Phi_G(1,2)$ in (7.28) and for $\Phi_E(1,2)$ in (7.30) can be performed by expanding the one-electron orbitals φ_s and φ_a in terms of the 1s orbital of the H atom using the confocal elliptic coordinates in Fig. 6.5. The solid curves in Fig. 7.3 show a crude sketch of the computational result.

As is seen from the figure, the experimental result is crudely approximated as was the case for the Heitler-London approximation shown in Fig. 7.2.

What is important here is not the comparison between the two approximations but the fact that the LCAO-MO approach is also capable of reproducing the essential feature of bonding in the H_2 molecule despite the crudeness of the approximation. This feature is very significant because the LCAO-MO approximation can be extended beyond the H_2 molecule to approximate many other molecules.

Although the result in Fig. 7.3 was obtained by using confocal elliptic coordinates, as in the case of the Heitler-London approach, in the LCAO-MO approximation for general polyatomic molecules rectangular coordinates (Cartesian coordinates) are used. Nowadays, any three dimensional molecule can be treated via Gaussian type basis functions utilizing Cartesian coordinate expressions, which has made the MO approximation both very powerful and versatile in the quantum mechanical analysis of molecules (cf. Sect. 8.2.1).

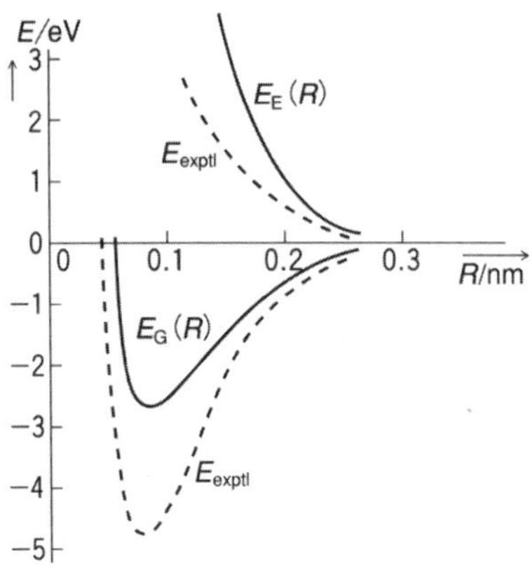

Fig. 7.3. Eigenvalues of the H_2 molecule in the ground and the first excited electronic states calculated with the LCAO-MO approximation (D.A. McQuarrie, J.D. Simon, 1997)

8 Polyatomic Molecules: Towards an Understanding of Chemical Bonds in Polyatomic Molecules

The central problem of the chemical bond in polyatomic molecules is to clarify the relationship between constituent atoms and the character of the bonds between them. This problem has been studied on the basis of the valence bond (VB) theory, a generalization of the Heitler-London theory, and on the basis of the molecular orbital (MO) theory. In this chapter we start by outlining the interelectronic interaction by way of the MO theory, which is currently popular. The computational techniques in molecular quantum mechanics have made remarkable progress since the 1970's. Nowadays, computations based on the first principle of quantum mechanics have reached such a level that computations, even though approximations are used by necessity, reproduce and predict experimental results quite satisfactorily in many cases. The use of computational techniques has also been greatly broadened to cover, e.g., complex biological molecules and molecular aggregates. In accordance with such progress, textbooks and manuals for practical computations are now available ubiquitously (e.g., W.J. Hehre et al.,1986; R.G. Parr and W. Yang, 1989; J.B. Foresman and Æ. Frisch, 1996; A. Szabo and N.S. Ostrand, 1996; F. Jensen, 1999; I.N. Levine, 2000). This book, however, is not intended to present such technicalities, but aims to provide a fundamental conceptual understanding of chemical bonding. Thus, in this chapter emphasis will be placed on the general tactics utilized for the analysis of the interelectronic interactions important in polyatomic molecules. The tactics are common to the analysis of multi-electron atoms (cf. Chap. 4).

After a general introduction to multi-electron systems, examples of the computational results for representative atoms and molecules will be presented for comparison with pertinent experimental data. Then, traditional concepts such as the covalent bond, the ionic bond, and the coordinate bond will be examined critically.

When molecules have geometrical symmetries in their nuclear arrangement, group theoretical analysis of the electronic structure becomes a powerful tool. If the equilibrium nuclear arrangement of such molecules is stationary or nearly so, the point group theory, whose group elements are symmetry operations on the nuclear geometry, is applicable. If the molecule under study is fluxional, i.e., the amplitude of the nuclear displacement is large as in the case of, e.g., van der Waals molecules, the permutation-inversion group theory is employed instead. Due to the limitations of space, however, the group theory as applied to the problem of chemical bonding will not be presented systematically in this book, however the results of group theoretical analyses will be discussed when appropriate.

At the end of this chapter, some prospects for future study of chemical bonding will be discussed.

8.1 The Schrödinger Equation for Multi-Electron Systems and its Approximate Solution
There are two major features in the problems of polyatomic molecules. Firstly, the molecules are multi-centered systems, and secondly, consideration of the interelectronic interaction is essential. The first step in considering the interaction is to follow the Hartree-Fock method. This method, first utilized in the study of multi-electron atoms, can be applied also to multi-centered polyatomic molecules. In the practical application, a method developed by Roothaan is employed. The Roothaan method gives approximate solutions in terms of self-consistent molecular orbitals (SCF-MO).

8.2 Examples of the SCF-MO Calculations and Comparisons with Experimental Results
Concrete examples of the Hartree-Fock-Roothaan method will be presented for some ten-electron systems. A relationship between the delocalized SCF-MO and the localized orbital as well as the hybrid orbital will also be discussed.

8.3 Critical Comments on the Classification of Chemical Bonds
Such conventional notions as the covalent bond, the ionic bond, and the coordinate bond are critically reviewed.

8.4 Some Topics and Prospects for the Study of Chemical Bonding
A summary and the prospects for future developments in the study of chemical bonding will be briefly discussed.

8.1 The Schrödinger Equation for Multi-Electron Systems and its Approximate Solution

In this section an approximate solution will be discussed for the Schrödinger equation of atoms and molecules with the interelectronic interaction explicitly taken into account.

8.1.1 Outline of the Hartree-Fock Method to Include the Interelectronic Interaction

The Schrödinger equation for a molecule is given by (5.1) which is reproduced below.

$$\hat{H}\Phi(\boldsymbol{r},\boldsymbol{R}) = E\Phi(\boldsymbol{r},\boldsymbol{R}) \tag{5.1}$$

$$\hat{H} = \hat{T}_{\mathrm{e}} + \hat{T}_{\mathrm{N}} + \hat{V}_{\mathrm{ee}} + \hat{V}_{\mathrm{eN}} + \hat{V}_{\mathrm{NN}} \tag{5.2}$$

$$\Phi(\boldsymbol{r},\boldsymbol{R}) \cong \Phi_{\mathrm{N}}(\boldsymbol{R})\Phi_{\mathrm{e}}(\boldsymbol{r},\boldsymbol{R}) \tag{5.4}$$

The symbols $\boldsymbol{r}$ and $\boldsymbol{R}$ represent the spatial coordinates of electrons and nuclei inclusively. The concrete forms of each term of the right-hand side of (5.2) are given by (5.3) in Chap. 5.

The Hamiltonian for multi-electron atoms, for which only computational results were summarized in Sect. 4.4, is similar to that in (5.2), with the last term $\hat{V}_{\mathrm{NN}}$ missing. In multi-electron atoms the term $\hat{T}_{\mathrm{N}}$ is also omissible because of the large mass of nuclei, so that the Hamiltonian for multi-electron atoms reduces to $\hat{H} = \hat{T}_{\mathrm{e}} + \hat{V}_{\mathrm{ee}} + \hat{V}_{\mathrm{eN}}$ as given in (4.21).

Returning to polyatomic molecules, it is seen that the electronic energy $E_{\mathrm{e}}(\boldsymbol{R}_{\mathrm{fixed}})$ and the electronic eigenfunction $\Phi_{\mathrm{e}}(\boldsymbol{r},\boldsymbol{R}_{\mathrm{fixed}})$ satisfy (5.10) where $\boldsymbol{R}_{\mathrm{fixed}}$ is regarded as a parameter and is simply rewritten as $\boldsymbol{R}$ hereafter.

$$\hat{H}_{\mathrm{e}}\Phi_{\mathrm{e}}(\boldsymbol{r},\boldsymbol{R}) = E_{\mathrm{e}}(\boldsymbol{R})\Phi_{\mathrm{e}}(\boldsymbol{r},\boldsymbol{R}) \tag{5.10}$$

In the following, the subscript e for $\Phi_{\mathrm{e}}(\boldsymbol{r},\boldsymbol{R})$ and the parameter $\boldsymbol{R}$ are omitted, for simplicity, and the eigenfunction is written as $\Phi_{\mathrm{e}}(\boldsymbol{r},\boldsymbol{R}) = \Phi(\boldsymbol{r}_1,\boldsymbol{r}_2,\dots) = \Phi(1,2,\dots)$. The explicit form of the Hamiltonian $\hat{H}_{\mathrm{e}}$ in (5.10) is given in (8.1) below.

$$\hat{H}_{\mathrm{e}} = \sum_i -\hbar^2\Delta_i/(2m_{\mathrm{e}}) - \sum_i\sum_P Z_P\frac{e^2}{4\pi\varepsilon_0 r_{iP}} + \sum_{i>j}\frac{e^2}{4\pi\varepsilon_0 r_{ij}} \tag{8.1}$$

In this section the nuclear arrangement is assumed to be fixed at its equilibrium position with no molecular vibration nor rotation, which means $\hat{T}_{\mathrm{N}} = 0$. Then, the overall Hamiltonian $\hat{H}$ of a molecule is written as follows.

$$\hat{H} = \hat{T}_{\mathrm{e}} + \hat{V}_{\mathrm{ee}} + \hat{V}_{\mathrm{eN}} + \hat{V}_{\mathrm{NN}} = \hat{H}_{\mathrm{e}} + \hat{V}_{\mathrm{NN}} \tag{8.2}$$

The term $\hat{H}_{\mathrm{e}}$ in (8.2) can be divided into the one-electron component $\hat{H}^0(i)$ and the interelectronic repulsive component $\hat{H}'(i,j)$ as follows.

$$\hat{H}_{\mathrm{e}} = \sum_i \hat{H}^0(i) + \sum_{i>j}\hat{H}'(i,j) \tag{8.3}$$

$$\hat{H}^0(i) = -\hbar^2\Delta_i/(2m_{\mathrm{e}}) - \sum_P Z_P\frac{e^2}{4\pi\varepsilon_0 r_{iP}} \tag{8.4}$$

$$\hat{H}'(i,j) = \sum_{i>j}\frac{e^2}{4\pi\varepsilon_0 r_{ij}} \tag{8.5}$$

As stated in Chap. 5, the term $\hat{V}_{\mathrm{NN}}$ in (5.2) and (5.3) or in (8.2) is merely a constant, $\sum_{P>Q} Z_P Z_Q e^2/(4\pi\varepsilon_0 R_{PQ,\mathrm{fixed}})$ at the equilibrium nuclear position, so that it will be denoted simply as V_{NN}.

The eigenfunction in (5.10) reproduced above must be anti-symmetric to obey the Pauli exclusion principle according to the statement in Sect. 4.1. The anti-symmetrization is achieved by expressing the eigenfunction $\Phi_e(\boldsymbol{r}, \boldsymbol{R}) \equiv \Phi(1, 2, \dots)$ with a Slater determinant using one-electron functions such as $\phi_{k_1}(\boldsymbol{r}_1\sigma_1) \equiv \phi_{k_1}(1)$, which is for electron 1 occupying the k_1-th spin orbital, and is supposed to be extended over the whole atom or the entire molecule under consideration. In the case of molecules, such a one-electron function is called generally the molecular orbital (MO). This function is a generalization of the LCAO-MO in Chaps. 6 and 7. In the general MO, no artificial restriction such as the linear combination of atomic orbitals is imposed.

Most molecules in the ground electronic state are in the singlet spin state. Molecules in such a spin state are the major subject in the study of chemical bonding in this book. The usual procedure to describe such a state is to assume that any two electrons share one and the same spatial function $\phi_{k_i}(\boldsymbol{r})$ but that they occupy different spin functions $\alpha(\sigma)$ and $\beta(\sigma)$. The assumption is a type of restriction upon the freedom of the electrons. In so far as this assumption is admitted, the eigenfunction in the form of the Slater determinant can be expressed compactly as in (8.6) below.

$$\Phi(1, 2, \dots, 2n) = \|\phi_{k_1}(1)\overline{\phi}_{k_1}(2)\phi_{k_2}(3)\overline{\phi}_{k_2}(4)\dots\phi_{k_n}(2n-1)\overline{\phi}_{k_n}(2n)\| \quad (8.6)$$

In (8.6), the spin-up and the spin-down functions $\alpha(\sigma)$ and $\beta(\sigma)$ are denoted as $\phi_{k_i}\alpha \equiv \phi_{k_i}$ and $\phi_{k_i}\beta \equiv \overline{\phi}_{k_i}$ together with the spatial component. The restriction imposed as in (8.6) is a significant artificial constraint on the freedom of the individual spin orbitals because there should be no a priori necessity for two electrons to share the same spatial component. Fortunately, however, it is known that most molecules with even numbers of electrons can be fairly well approximated by (8.6), especially in the vicinity of their stable geometrical structure, provided that the spatial component $\phi_{k_i}(\boldsymbol{r})$ has been 'properly' prepared. Thus, the problem is how to prepare or determine the orbital $\phi_{k_i}(\boldsymbol{r})$ properly. The Hartree-Fock method briefly introduced in Sect. 4.4 was conceived to meet this very requirement. In the following section the Hartree-Fock method will be explored. Strictly speaking, the method should be called the Restricted-Hartree- Fock method (RHF) in the sense that two electrons are artificially restricted to one and the same spatial component of the spin orbitals. As a less restrictive approximation there is the Unrestricted-Hartree-Fock method (UHF), but in this book we will concentrate on the RHF method because the RHF approximation suffices for the majority of cases we need be concerned with. Before proceeding to the explanation of the Hartree-Fock method, a brief comment on the electron correlation will be made here.

As stated in connection with (4.22), the one-electron function sought, $\phi_i(1)$, is obtained by the Hartree-Fock equation which assumes that electron 1 is in a potential where any effect due to the other electrons is averaged. This assumption is tantamount to the negation of any correlation between specific pairs of electrons, and thus the Hartree-Fock method is incapable of accounting for any correlation between pairs of electrons. The problem posed by this electron correlation is one of the major nuisances in quantum chemistry, especially in molecules with geometrical arrangements far from the equilibrium. As an example of such a molecule, one may consider the reaction intermediates, important in chemical reactions. Even in the vicinity of equilibrium arrangements, the problem of correlation becomes serious if the required level of computational accuracy is high. To overcome this problem of electron correlation, various techniques have been developed such as the adoption of multi-electron functions which take into account the electron correlation. Such topics, however, are beyond the scope of this book.

We now return to the procedure to determine the 'proper' MO mentioned above. The technique to be used here is the calculus of variation, the fundamentals of which have been explained in Sects. 2.1.15 and 2.2.8. In the present problem, the expectation value of energy corresponding to (2.55) is written as follows, under the assumption that the molecular orbitals sought are orthonormalized.

$$E = \int \cdots \int \Phi^*(1,2,\ldots,2n)\hat{H}\Phi(1,2,\ldots,2n)\mathrm{d}\tau_1\mathrm{d}\tau_2\ldots\mathrm{d}\tau_{2n}, \tag{8.7}$$

In (8.7) the volume elements, such as $\mathrm{d}\tau_1$, represent the product of the spatial component and the electron spin component, i.e., $\mathrm{d}\tau_1 \equiv \mathrm{d}v_1\mathrm{d}\sigma_1$. So far the k_i-th spin orbital was denoted as $\phi_{k_i}(\boldsymbol{r})\alpha(\sigma)$ or $\phi_{k_i}(\boldsymbol{r})\beta(\sigma)$, but hereafter the notation k_i will be simplified as i, and all the spatial functions for electron 1 will be denoted as $\phi_i(\boldsymbol{r}) \equiv \phi_i(1)$ or $\phi_j(\boldsymbol{r}) \equiv \phi_j(1)$, which are assumed to have been orthonormalized. The orthogonality ensures that each one-electron function is linearly independent as explained in Sects. 2.1.9, 2.1.13, and 2.1.14. We assume this orthogonality prior to the determination of the concrete form of the function. This assumption can be expressed as follows:

$$\int \phi_i^*(1)\phi_j(1)\mathrm{d}v_1 = \delta_{ij} \tag{8.8}$$

In the calculus of variation, we search for a one-electron function which satisfies the relation $\delta E = 0$ when the function for electron 1, e.g., is displaced as $\phi_i(1) \to \phi_i(1) + \delta\phi_i(1)$. This is a necessary condition to find the minimum of E.

In the present case, this search must be carried out under the restriction of (8.8). A conventional technique, in such a conditional calculus of variation, is the Lagrange undetermined multiplier method*. In this book only the result of the use of this technique will be presented as in (8.9). A detailed derivation of (8.9) can be found in an instructive article by C.C. Roothaan (1951), which is highly recommended as a must-read for motivated readers.

$$\left[\hat{H}^0(1) + \sum_{j=1}^{n}(2\hat{J}_j(1) - \hat{K}_j(1))\right]\phi_i(1) = \varepsilon_i\phi_i(1) \tag{8.9}$$

At present, only a summary of the above equation will be presented. Understanding the function of operators $\hat{J}_j$ and $\hat{K}_j$ in (8.9) should lead the reader to grasp how to treat the problem of multi-electron systems.

As will be shown shortly (cf. (8.11) and (8.12)), (8.9) implies that the sought one-electron function $\phi_i(1)$ is the solution of that equation. Before examining the meaning of each term in (8.9), a compact expression of (8.9) is introduced as in (8.10) below, where the symbol $\hat{F}(1)$ is named the Fock operator and the equation is called the (Hartree-)Fock equation or the Euler equation in the calculus of variation.

$$\hat{F}(1)\phi_i(1) = \varepsilon_i \phi_i(1) \tag{8.10}$$

Looking at each term of (8.9), it is found that the first term represents the Hamiltonian for electron 1 when it is in the field of the bare nuclei (cf. (8.4)). As for the second term summed over $j = 1$ through n, it represents the effect of all the other electrons upon electron 1. This term originates from the operator in (8.5), which represents the interelectronic repulsion. These conclusions will be confirmed after the examination of the meaning of the operators $\hat{J}_j(1)$ and $\hat{K}_j(1)$, which follows.

The two operators are defined by (8.11) and (8.12) below, and are called the Coulomb *operator* and the exchange *operator*, respectively. The italics are to draw attention to the fact that they are operators, which should be clearly differentiated from the *integrals* bearing similar adjectives to appear later in (8.17) and (8.18).

$$\hat{J}_j(1)\phi_i(1) \equiv \left[\frac{e^2}{4\pi\varepsilon_0}\int \frac{\phi_j^*(2)\phi_j(2)\mathrm{d}\tau_2}{r_{12}}\right]\phi_i(1) \tag{8.11}$$

$$\hat{K}_j(1)\phi_i(1) \equiv \left[\frac{e^2}{4\pi\varepsilon_0}\int \frac{\phi_j^*(2)\phi_i(2)\mathrm{d}\tau_2}{r_{12}}\right]\phi_j(1) \tag{8.12}$$

As stated in Sect. 1.2, operators in general, are defined independently of the functions to be operated upon. However, in the above cases for $\hat{J}_j(1)$ and $\hat{K}_j(1)$, they are defined by involving the functions operated upon in rather complicated manners. The reason for these particular definitions is that the Coulomb integral and the exchange integral in (8.17) and (8.18) can be expressed compactly in terms of the operators $\hat{J}_j(1)$ and $\hat{K}_j(1)$ as defined in (8.11) and (8.12).

The operator $\hat{J}_j(1)$ may easily be seen as representing the potential for electron 1 in the orbital ϕ_i, which is under the influence of the Coulombic repulsion by electron 2 occupying the orbital ϕ_j. In contrast, such a straightforward interpretation is not possible for the operator $\hat{K}_j(1)$ because (8.12) requires the relocation of electron 1 from ϕ_i to ϕ_j and the occupation of both orbitals by electron 2. This operator $\hat{K}_j(1)$ is derived automatically from the requirement that the exchange of electrons 1 and 2 in the Slater determinant of (8.6) must reverse the sign of the determinant due to the Pauli exclusion principle. It is noticed that the operator $\hat{J}_j(1)$ remains the same, whether the electron spin components to be associated with the spatial components $\phi_i(1)$ and $\phi_j(2)$ are alike as $\alpha(1)\alpha(2)$ or as $\beta(1)\beta(2)$, or they are different as $\alpha(1)\beta(2)$ or as $\beta(1)\alpha(2)$. In contrast, the operator $\hat{K}_j(1)$ survives only when the spin components of both electrons are the same. Otherwise, the integral in (8.12) becomes 0 due to the orthonormality of the spin function. In other words, the interelectronic repulsion represented by the summation of the $\hat{J}_j(1)$ terms on the left-hand side of (8.9) is mitigated by an amount of $-\sum_{j=1}^{n} \hat{K}_j(1)$ if the spin states of the interacting electrons are the same. Alternatively, the effect of the operator $\hat{K}_j(1)$ in (8.12) may be regarded as if virtual positive charges were produced to mitigate the negative-to-negative interelectronic repulsion. Such a virtual positive charge may sometimes be called the Fermi positive hole.

As stated above, (8.9) (or (8.10)) describes electron 1 in the field of the bare nuclei and the averaged field due to all the other electrons. Thus, if the terms $\hat{J}_j(1)$ and $\hat{K}_j(1)$ were missing in (8.9), electron 1 would be exposed exclusively to the field of the nuclei. Under such an extreme condition, we have the relation below where $\hat{H}^0(1)$ is the same as $\hat{H}^0(i)$ in (8.4) and ε_i in (8.9) is replaced with ε_i^0. The notations in (8.13) will be used later in (8.15) and (8.16).

$$\varepsilon_i^0 = \int \phi_i^*(1)\hat{H}^0(1)\phi_i(1)\mathrm{d}\tau_1 \equiv H_{ii}^0 \tag{8.13}$$

The argument made so far is applicable not only to polyatomic molecules but also to multi-electron atoms; in the latter systems the term $\hat{V}_{\mathrm{NN}}$ in (8.2) is deleted. In multi-electron atoms, electron 1 obeying (8.9) is regarded as being in a spherical potential because the averaging of the effects of all the other electrons is equivalent to approximating the potential as an isotropic central field.

Although nothing has been mentioned as to ε_i in (8.9), it was introduced as a Lagrangean undetermined multiplier in the cited paper by Roothaan. A physical meaning can be endowed now to this multiplier; it is the energy eigenvalue for an electron in the averaged field in both polyatomic molecules and multi-electron atoms. This energy is called the SCF canonical orbital energy after the SCF method to be explained shortly.

Since ε_i is the energy eigenvalue of the Hermitian operator $\hat{F}(1)$, it must be a real number. To be more concrete, it is an energy approximately equal to that required to ionize electron 1 from the one-electron orbital $\phi_i(1)$ of a multi-electron atom or a polyatomic molecule. Ionization here means the removal of electron 1 with zero kinetic energy to an infinite separation from the remaining positive ion. The energy necessary for this ionization is called the ionization energy E_{i} so that the above statement can be expressed as $\varepsilon_i \cong -E_{\mathrm{i}} < 0$ (cf. Sects. 4.4 and 8.2.3). This relation is called Koopmans' theorem and was introduced briefly in Sect. 4.4.

As to the procedure for solving (8.9), a general approach will be given subsequently in Sect. 8.1.2. As a preliminary step, the integrals related to the Hartree-Fock equation are explained first. Let us assume that the one-electron function $\phi_i(1)$ in (8.9) has been found by some means. Then, the expectation value E in (8.7) can be obtained by substituting the found one-electron functions into (8.6) and (8.7). The final expression for E turns out to be as in (8.14) below with the abbreviations in (8.15) through (8.18):

$$E = 2\sum_{i=1}^{n} \varepsilon_i - \sum_{i=1}^{n}\sum_{j=1}^{n}(2J_{ij} - K_{ij}) + V_{\mathrm{NN}} \tag{8.14}$$

The last term on the right-hand side of (8.14) represents a constant internuclear repulsion energy which appears in polyatomic molecules but not in multi-electron atoms (cf. (8.2)). The remaining terms are all definite integrals with regard to the operators in (8.9) through (8.12), the results being as follows:

$$\varepsilon_i = H_{ii}^0 + \sum_{j=1}^{n}(2J_{ij} - K_{ij}) \tag{8.15}$$

$$H_{ii}^0 = \int \phi_i^*(1)\hat{H}^0(1)\phi_i(1)\mathrm{d}\tau_1 \equiv \varepsilon_i^0 \tag{8.16}$$

$$\begin{aligned} J_{ij} &= \frac{e^2}{4\pi\varepsilon_0}\int \phi_i^*(1)\hat{J}_j\phi_i(1)\mathrm{d}\tau_1 \\ &= \frac{e^2}{4\pi\varepsilon_0}\iint \phi_i^*(1)\phi_i(1)\frac{1}{r_{12}}\phi_j^*(2)\phi_j(2)\mathrm{d}\tau_1\mathrm{d}\tau_2 \end{aligned} \tag{8.17}$$

$$\begin{aligned} K_{ij} &= \frac{e^2}{4\pi\varepsilon_0}\int \phi_i^*(1)\hat{K}_j\phi_i(1)\mathrm{d}\tau_1 \\ &= \frac{e^2}{4\pi\varepsilon_0}\iint \phi_i^*(1)\phi_j(1)\frac{1}{r_{12}}\phi_j^*(2)\phi_i(2)\mathrm{d}\tau_1\mathrm{d}\tau_2 \end{aligned} \tag{8.18}$$

The definite integrals in (8.17) and (8.18) have the dimension of energy and are called the Coulomb integral and the exchange integral, respectively, in correspondence to the operators $\hat{J}_j$ and $\hat{K}_j$ in (8.11) and (8.12).

It is noted that the energy E in (8.14) is not equal to a simple sum of the orbital energies ε_i. This is because such a simple sum incorrectly doubly counts the interelectronic interaction energy appearing as the second term in (8.15). Additionally, the total energy E may be written also as follows by substituting (8.15) into (8.14).

$$E = 2\sum_{i=1}^{n} H_{ii}^0 + \sum_{i=1}^{n}\sum_{j=1}^{n}(2J_{ij} - K_{ij}) + V_{\mathrm{NN}} \tag{8.19}$$

The definite integral $K_{ij}(> 0)$ survives only when electrons 1 and 2 are in the same spin state, as previously commented in connection with the operator $\hat{K}_j(1)$ in (8.9). Since K_{ij} in (8.19) bears a minus sign, the surviving term contributes to the lowering (stabilization) of the total energy.

The total energy E in (8.7) is to be obtained by a $6n$-fold integral because the total $2n$ electrons have three degrees of freedom in their spatial coordinates. However, (8.10) through (8.18) show that the $6n$-fold integral is actually reduced to a 3-fold integral (H_{ii}^0) for electron 1 and 6-fold integrals (J_{ij}, K_{ij}) for electrons 1 and 2. This is a great saving of labor in the actual computation.

Superficially, (8.10) appears to be an eigenvalue equation to give the eigenfunction $\phi_i(1)$ and the eigenvalue ε_i. However, since the components of the operator $\hat{F}(1)$, i.e., $\hat{J}_j(1)$ and $\hat{K}_j(1)$ are defined by (8.11) and (8.12), the concrete form of $\hat{F}(1)$ cannot be fixed unless the solution $\phi_i(1)$, to be determined by (8.10), is given somehow. Hartree and Fock have overcome this dilemma originating from the interelectronic interaction. An outline of their maneuver follows.

First, a set of molecular orbitals $\{\phi_i^0,\ i = 1, 2, \ldots, n\}$ is put forth as an initial guess. Then, using this set, the operators in (8.11) and (8.12) are constructed to yield the Fock operator $\hat{F}(1)$ in (8.10). By solving this equation a new set of molecular orbitals $\{\phi_i^{(1)},\ i = 1, 2, \ldots, n\}$ is obtained. With these new orbitals a renewed Fock operator is obtained referring to (8.11) and (8.12). The above iterative procedure is repeated until the solution of the last Fock operator in the cycle converges to the preceding solution within some tolerable allowance. The last solution of the iteration may be said to be self-consistent so that the procedure is called the self-consistent field (SCF) method, or equivalently the Hartree-Fock method.

As S. Huzinaga (1980) stated, however, it is not at all easy to actually determine the SCF orbital, contrary to the apparently simple scenario stated above. After all, the estimation of the interelectronic interaction in multi-electron systems is quite a demanding task.

Because of this difficulty in the practical application of the general Hartree-Fock method, a more workable method was needed, and the Hartree-Fock-Roothaan method was invented, which will be explained in the next section. Before proceeding, an important comment on the SCF theory will be made; as stated in connection with (8.8), the one-electron orbitals in the SCF theory are mutually orthogonalized. This is a big difference from the valence bond (VB) theory, which was initiated by Heitler and London and developed by such people as Pauling and Slater. In the VB theory, the orthogonality is not guaranteed automatically, in contrast to the SCF-MO theory. The overwhelming prevalence of the SCF-MO theory versus the VB theory in modern quantum chemistry owes a great deal to this difference among others.

8.1.2 A Practical Solution to the Problem of Interelectronic Interaction: The Hartree-Fock-Roothaan Method

As stated in Sect. 8.1.1, the Fock operator $\hat{F}(1)$ in (8.10) is defined by the one-electron function $\phi_i(1)$, which is obtained via a procedure best characterized as 'going around in circles'. Furthermore, the operator involves a cumbersome application of the Laplacian appearing in $\hat{H}^0(1)$, which makes the computation laborious. In order to circumvent these problems, a more facile method was developed by Roothaan (1951). In this method a trial function $\varphi_i(1)$ to approximate the sought for function $\phi_i(1)$ is assumed to be a linear combination of finite known atomic orbitals (AO), or more generally, some basis functions $\chi_\mu(1)$ with the total number equal to M, as follows.

$$\varphi_i(1) = \sum_{\mu=1}^{M} \chi_\mu(1) c_{\mu i} \tag{8.20}$$

The above trial function $\varphi_i(1)$ of an LCAO-MO type approximates the orbital of electron 1 delocalized over the whole molecule. The Greek suffixes μ and ν are for the AO (or the basis functions) and the Roman suffixes are to differentiate the one-electron functions $\phi_i(1)$, which are replaced here with the trial functions $\varphi_i(1)$. The important orthogonality condition in (8.8) is, then, written as follows.

$$\begin{aligned} \int \varphi_i^*(1)\varphi_j(1)\mathrm{d}\tau_1 &= \sum_\mu \sum_\nu c_{\mu i}^* c_{\nu j} \int \chi_\mu^*(1)\chi_\nu(1)\mathrm{d}\tau_1 \\ &= \sum_\mu \sum_\nu c_{\mu i}^* c_{\nu j} S_{\mu\nu} = \delta_{ij} \end{aligned} \tag{8.21}$$

$$\int \chi_\mu^*(1)\chi_\nu(1)\mathrm{d}\tau_1 \equiv S_{\mu\nu} \tag{8.22}$$

With this restriction imposed on $\varphi_i(1)$ the accuracy of $\varphi_i(1)$ can be improved by increasing the number of basis functions in (8.20). In principle, basis functions having the completeness property should be used for the expansion of (8.20) (cf. the argument on completeness in Sect. 2.1.10). If such a complete set were found by luck somehow, then, the 'exact Hartree-Fock solution' would be in our hand. However, since such luck is obviously unrealistic, we have to make do with some reasonably good, but incomplete basis functions.

By the way, chemists are aware by experience that atoms constituting molecules keep their characters in various aspects to a remarkable extent. Therefore, in the purely mathematical procedure of solving the Fock equation in (8.10) also, we anticipate that orbitals to describe molecules can be approximated by a suitable set of orbitals to describe atoms. For this reason chemists attempt to express molecular orbitals by a finite set of basis functions to describe atoms. These are centered on the position of each atom in the molecule and are sufficient in approximating the atoms, leading to good transferability among various types of molecules. Such a set of basis basis functions can be regarded as proper to each atom and usable commonly in the computation of various different molecules.

As an example, consider the H_2 molecule in Chap. 7. There, the molecular orbital was expressed by the linear combination of 1s atomic orbitals exclusively. However, as remarked in Sect. 7.4, restricting to just the 1s AO cannot describe properly the distortion and expansion of the electronic cloud in the molecule, which become prominent especially at short internuclear distances. To remove such a deficiency we replace the simple 1s AO with a sum of more than one 1s-type functions whose coefficients in the exponent are varied from the fixed value in Table 3.1. Such a sum of 1s-type functions can describe flexibly a wider range of space around the nuclear center. The smaller the coefficient is, the wider the space covered. If p_x-type functions in Table 3.2 are included in addition to the above 1s-type functions on both H atoms lying in the x axis, they describe effectively the distortion caused by the counterpart nuclei. By now, plenty of such sets of atomic basis functions are catalogued to meet the various demands of problems over a wide range of accuracy requirements for computational solutions. However, it must be admitted that with an increase in the number of basis functions, the intuitively sensible result of using the simple LCAO-MO becomes less and less clear. As a compromise, an outline of the LCAO-MO type computation with the minimal basis set or the minimum basis set will be introduced here. The minimal basis set consists of all the atomic orbitals occupied by electrons in the ground electronic state of the atom plus the unoccupied orbitals in the subshell (nl), where n is the principal quantum number and l is the azimuthal quantum number (cf. Sect. 4.2 and Table 4.1).

As a concrete example, the case of HLi will be explained. In the H atom the minimal basis set includes only the 1s while in the Li atom the set consists of the occupied 1s and 2s orbitals plus unoccupied orbitals of $n = 2$ and $l = 1$, i.e., $2\mathrm{p}_x$, $2\mathrm{p}_y$, and $2\mathrm{p}_z$.

In Roothaan's method, the coefficient $c_{\mu i}$ in (8.20) is determined by Ritz's calculus of variation, irrespective of the basis set adopted. That is to say, those coefficients that give $\delta E = 0$ with the displacement of $c_{\mu i}$ to $c_{\mu i} + \delta c_{\mu i}$ are looked for. The practical procedure is as follows; in place of the general one-electron function $\phi_i(1)$ in Sect. 8.1.1, the trial function $\varphi_i(1)$ in (8.20) is substituted to (8.9) to obtain the following simultaneous equations for the coefficient $c_{\nu i}$.

$$\sum_{\nu=1}^{M} c_{\nu i}\hat{F}(1)\chi_\nu(1) = \varepsilon_i \sum_{\nu=1}^{M} c_{\nu i}\chi_\nu(1) \tag{8.23}$$

Multiplication by $\chi_\mu^*(1)$ from the left on both sides of (8.23) and integration with regard to the spatial coordinates of electron 1 leads to the following M equations with the identities $F_{\mu\nu} \equiv \int \chi_\mu^*(1)\hat{F}(1)\chi_\nu(1)\mathrm{d}\tau_1$ and $S_{\mu\nu} \equiv \int \chi_\mu^*(1)\chi_\nu(1)\mathrm{d}\tau_1$.

$$\sum_{\nu=1}^{M} c_{\nu i}(F_{\mu\nu} - \varepsilon_i S_{\mu\nu}) = 0, \quad \mu, \nu = 1, 2, \ldots, M \tag{8.24}$$

This equation is called the Hartree-Fock-Roothaan equation. Using the notations $\varphi_i = \sum_\mu \chi_\mu c_{\mu i}$, $\varphi_i = \sum_\nu \chi_\nu c_{\nu i}$, $\varphi_j = \sum_\lambda \chi_\lambda c_{\lambda j}$ and $\varphi_j = \sum_\sigma \chi_\sigma c_{\sigma j}$, the Hartree-Fock-Roothaan equation is explicitly written as in (8.25) below with the abbreviation of $H_{\mu\nu}^0 \equiv \int \chi_\mu^*(1)\hat{H}^0(1)\chi_\nu(1)\mathrm{d}\tau_1$.

$$\begin{aligned}
&\sum_\nu c_{\nu i} H_{\mu\nu}^0 \\
&+ \frac{e^2}{4\pi\varepsilon_0}\sum_\nu c_{\nu i}\sum_j^{occ}\sum_\lambda\sum_\sigma c_{\lambda j}^* c_{\sigma j}\iint\left\{\chi_\mu^*(1)\chi_\nu(1)\frac{2}{r_{12}}\chi_\lambda^*(2)\chi_\sigma(2)\right\}\mathrm{d}\tau_1\mathrm{d}\tau_2 \\
&- \frac{e^2}{4\pi\varepsilon_0}\sum_\nu c_{\nu i}\sum_j^{occ}\sum_\lambda\sum_\sigma c_{\lambda j}^* c_{\sigma j}\iint\left\{\chi_\mu^*(1)\chi_\sigma(1)\frac{1}{r_{12}}\chi_\lambda^*(2)\chi_\nu(2)\right\}\mathrm{d}\tau_1\mathrm{d}\tau_2 \\
&- \sum_\nu \varepsilon_i c_{\nu i} S_{\mu\nu} = 0
\end{aligned} \tag{8.25}$$

In the above the notation *occ* means that the sum is over all the n trial one-electron functions occupied by $2n$ electrons in all. The above equation appears complicated, but it should be clear that the terms involving the factors shown in the round brackets {} represent the interelectronic interaction. Of course, these factors having the dimension of energy are constants as a result of the definite integrals with regard to the given atomic orbitals χ's. For the simultaneous equations of (8.24) or (8.25) to have significant solutions for the coefficients of the linear combination, the following secular equation (8.26) must hold as usual, where the notation det is the abbreviation of the determinant whose elements consist of the factors to be multiplied by the coefficients $c_{\nu i}$ which are sought.

$$\det(F_{\mu\nu} - \varepsilon_i S_{\mu\nu}) = 0 \tag{8.26}$$

This is a polynomial with regard to ε_i whose roots are the orbital energies. However, since the matrix element $F_{\mu\nu}$ contains the coefficients of the linear combination of the trial functions, i.e., $c_{\lambda i}$ and $c_{\sigma j}$ to be summed as in (8.25), it is necessary to resort to a procedure such as stated in Sect. 8.1.1. In practice, the initial set of coefficients must be prepared by some means. A drastically simple choice of an initial set is to ignore the interelectronic repulsive terms in (8.25), and one starts with the solution of the following simultaneous equations.

$$\sum_{\nu} \left[H^0_{\mu\nu} - \varepsilon^0_i S_{\mu\nu}\right] c_{\nu i} = 0, \ \mu = 1, 2, \ldots, M \tag{8.27}$$

The above equation determines one-electron orbitals $\varphi^0_i(1) = \sum_\nu \chi_\nu(1) c^0_{\nu i}$ for electrons delocalized over the whole molecule where the electrons feel the potential field of the bare nuclei. Once the coefficients $c^0_{\nu i}$ and the energies ε^0_i are determined, the iterative procedure for the coefficients and the energies is carried out until the desired convergence is attained. The converged energy ε_i and the one-electron orbital $\varphi_i(1) = \sum_\nu \chi_\nu(1) c_{\nu i}$ are called the SCF molecular orbital energy and the (Hartree-Fock) SCF canonical molecular orbital, respectively. For simplicity, hereafter the orbital will be called the canonical molecular orbital. A vector $\boldsymbol{C}_i \equiv (c_{1i}, c_{2i}, \ldots, c_{Mi})$ with the components of the coefficients for a given canonical molecular orbital, is called an eigenvector because this vector corresponds to the eigenvector in the general matrix eigenvalue problem in (8.28).

$$\boldsymbol{F}\boldsymbol{C}_i = \boldsymbol{S}\boldsymbol{C}_i \varepsilon_i \tag{8.28}$$

The matrix elements in $\boldsymbol{F}$ and $\boldsymbol{S}$ consist of the elements $F_{\mu\nu}$ and $S_{\mu\nu}$ in (8.26).

8.2 Examples of SCF-MO Calculations and Comparisons with Experimental Results

Molecular orbital computations based on the first principle of quantum mechanics, or more broadly speaking, quantum chemical computations are making rapid progress at present. It is becoming possible to reproduce experimental results by computations with 'chemical accuracy' and to predict experimental results. Chemical accuracy here means that the discrepancy between computation and experiment is roughly less than 0.1 eV for the energy change associated with various chemical events. To achieve accuracy to this extent it is necessary to increase the number of basis functions discussed in Sect. 8.1.2 and to take into account the electron correlation mentioned in Sect. 8.1.1.

These subjects, as mentioned already, are beyond the scope of the present book, however they are treated in many excellent publications, such as those introduced in the beginning of Chap. 8.

In this section the simplest version of the SCF-MO method introduced in Sect. 8.1 will be applied to some ten-electron atoms and molecules and the computational result will be examined in comparison with related experimental results. The regression to the simple method is deliberate because the aim of this book is to keep an intuitively clear picture of chemical bonding as much as possible, as stated in Sect. 8.2.2.

8.2.1 Computation for HF

In the following I will introduce the results of a computation done on HF in the early days by B.F. Ransil (1960), who used the minimal basis set and the Hartree-Fock-Roothaan approximation. A slight change by J.A. Pople and D.L. Beverage (1970) has been incorporated to Ransil's result.

First, the internuclear distance R between H and F is fixed to the experimental value of $R = 0.0917\,\text{nm}$. Next, in order to express the trial function φ in (8.20) by a linear combination of basis functions we adopt Slater-Type Orbitals (STO, χ_S) instead of atomic orbitals themselves. An STO is obtained by replacing the radial part $R_{n,l}(r)$ (cf. Table 3.1) in the total eigenfunction of one-electron atoms, $\phi(\boldsymbol{r}) = R_{n,l}(r)Y_{l,m}(\theta,\varphi)$, with a function of $R'_{n,l}(r) \equiv Nr^{n-1}\text{e}^{-\zeta r}$ as follows.

$$\chi_\text{S} = R'_{n,l}(r)Y_{l,m}(\theta,\varphi) \tag{8.29}$$

The constants N and ζ in $R'_{n,l}(r)$ are the normalization constant and an empirical parameter, respectively. In contrast to $R_{n,l}(r)$, $R'_{n,l}(r)$ has no node in the radial direction.

The STO function is an empirical basis function determined by Slater through analysis of various experimental data. The criteria used in determining the empirical parameters ζ were, first, that the function reproduces a significant amount of atomic spectroscopic experimental data and secondly that the computational difficulties are decreased in comparison with those of computations using the exact eigenfunctions of one-electron atoms. For this reason the STO can be regarded as mimicking the AOs with much easier functionality. Since we adopt STOs instead of AOs, we may call the trial function here as LCSTO-MO rather than LCAO-MO.

Before proceeding to the actual computation for HF on the basis of STOs, I will make a comment on another type of basis function called Gaussian-Type Orbitals (GTO, χ_G).

The remarkable progress in quantum chemical computations in the last quarter of the twentieth century is due to the use of GTOs rather than the use of STOs. The GTO first proposed by S.F. Boys (1950) can be written in terms of either the polar or the Cartesian coordinate system as in (8.30) below, where the indexes l_x, l_y, and l_z in the Cartesian expression are integers such as 0, 1, and 2, and the sum of l_x, l_y, and l_z determines the type of orbital. For example, $l_x + l_y + l_z = 1$ is a p-type orbital, and N is the normalization constant.

$$\begin{aligned}\chi_\text{G}(n,l,m,\zeta) &= Nr^{2n-l-2}\text{e}^{-\zeta r^2}Y_{l,m}(\theta,\varphi),\\ \chi_\text{G}(l_x,l_y,l_z,\zeta) &= Nx^{l_x}y^{l_y}z^{l_z}\text{e}^{-\zeta r^2}\end{aligned} \tag{8.30}$$

The indexes l_x, l_y, and l_z are not quantum numbers while the indexes, l, and m appearing in the polar coordinate expression retain the meaning of quantum numbers. In both expressions the parameter ζ in the exponent determines the degree of the dependence on r. This dependence of the Gaussian type makes the GTO inferior to the STO in two aspects. At the nucleus the GTO has 0 slope due to the r^2 dependence in the exponent, whereas the STO has a 'cusp' (discontinuous derivative of the radial part with regard to r) so that the GTO fails to represent the proper behavior of atomic orbitals near the nucleus. The other problem inherent in the GTO is that the radial component falls off too rapidly with r as compared with the STO.

In compensation for such inferiority, the GTO is superior with respect to computational efficiency; as is well known, the product of two Gaussian functions centered at two different points yields a new single Gaussian function centered at a certain point between the two. Thus, all three- and four-center two-electron repulsion integrals based on GTOs, which appear in computations of interelectronic interactions in multi-electron systems, can be reduced to two-center integrals, and thus, the computation time can be greatly reduced.

Furthermore, through the use of GTOs it is possible to compute directly the first and the second derivatives of energy with respect to geometric physical variables such as bond lengths and bond angles. This automatic computational functionality allows for an easy search for the equilibrium molecular geometry. Thus, although more GTOs are needed to achieve an accuracy comparable with that attained by using STOs, the increase in the number of GTOs is more than compensated for by the ease by which the required integrals can be computed. Therefore, in terms of computational efficiency, the GTO is by far superior to the STO.

A compromise which utilizes the strong points of both STOs and GTOs is to express an STO as the sum of several GTOs whose exponents have been determined by a best fit to reproduce the STO. Since STOs themselves are an approximation for AOs in molecules, however, nowadays, MOs are expressed directly by linear combinations of a number of GTOs optimized to reproduce atomic data.

Despite the above mentioned superiority of GTOs to the STOs in computation, the analysis of HF by Ransil which follows will be carried out in terms of STOs. These STOs mimic AOs minimally and provides simpler visualization of molecules in relation to the constituent atoms, which is in the spirit of this book aiming a fundamental understanding of chemical bonding.

We now proceed to Ransil's computation of HF in terms of the STOs which are listed below. These basis functions correspond to those in the first row of Table 8.1 for HF.

$$\chi_{\mathrm{F1s}} = \sqrt{8.7^3/\pi}\,\exp(-8.7r), \qquad \chi_{\mathrm{F2s}} = \sqrt{2.6^5/(3\pi)}\,r\exp(-2.6r),$$
$$\chi_{\mathrm{F2p}_x} = \sqrt{2.6^5/\pi}\,x\exp(-2.6r), \qquad \chi_{\mathrm{F2p}_y} = \sqrt{2.6^5/\pi}\,y\exp(-2.6r),$$
$$\chi_{\mathrm{F2p}_z} = \sqrt{2.6^5/\pi}\,z\exp(-2.6r), \qquad \chi_{\mathrm{H1}s} = \sqrt{1/\pi}\,\exp(-r)$$

In Table 8.1 below, the units of energy and distance are the atomic units (au). The au for energy is also called the hartree and the symbol E_{h} is used.

The conversion relations are $1E_\mathrm{h} = 4.3597\cdots \times 10^{-18}\,\mathrm{J} = 27.211\ldots\mathrm{eV}$. As for the au for distance, the Bohr radius a_0 is used which has a relation $1a_0 = 5.291\cdots \times 10^{-11}\,\mathrm{m}$.

Once the basis functions are given explicitly as above, the substantial work to follow is to determine the coefficients for the linear combination in (8.20) according to the procedure given in Sect. 8.1.2. The first step is to compute all the overlap integrals among the STOs. The next is to obtain $H^0_{\mu\nu}$ which can be computed straightforwardly because the operator $\hat{H}^0(1)$ is given explicitly in (8.4). Cumbersome computations are the double integrals for the interelectronic interaction appearing in (8.25), i.e., $\iint \chi^*_\mu(1)\chi_\nu(1)(1/r_{12})\chi^*_\lambda(2)\chi_\sigma(2)\mathrm{d}\tau_1\mathrm{d}\tau_2$ and $\iint \chi^*_\mu(1)\chi_\sigma(1)(1/r_{12})\chi^*_\lambda\cdot(2)\chi_\nu\mathrm{d}\tau_1\mathrm{d}\tau_2$, but these are obtained by standard numerical programs.

Up to this point all the necessary integrals are in hand and the next step is to solve (8.25) which is essentially a third order polynomial with regard to the coefficients c. This equation cannot be solved straightforwardly as stated in Sect. 8.1.2. Therefore, we have to start with some initial clue, such as (8.27), and repeat the solution of (8.25) and (8.26) to obtain coefficients which have sufficiently converged for the linear combination and the associated orbital energies.

The final converged results for HF are shown in the upper left of Table 8.1. The converged trial functions correspond to the desired LCAO-MO, i.e., $\varphi_i(1)$ in (8.20). The symbols such as $1\sigma, 2\sigma, \ldots$ in Table 8.1 are to differentiate each $\varphi_i(1)$. The meaning of these symbols will be explained subsequently in Sect. 8.2.2. The first row of Table 8.1 shows the minimal basis set, $\chi_\mu(1)$, corresponding to the basis functions in (8.20). The numbers in the main part of the table are the coefficients $c_{\mu i}$ in (8.20). In consideration of the inherent accuracy of the computation, data are given to two decimals places only. A more accurate result for CH_4 will be given later (cf. (8.62) through (8.66)). As stated in Sect. 8.1.2, the numbers in the table represent the vectorial components when the trial function in (8.20) is regarded as an eigenvector. Similarly, the values under the heading of ε_i in Table 8.1 correspond to the orbital energies, which correspond to the eigenvalues of (8.9) (or (8.10)).

In reference to Koopmans' theorem in Sect. 8.1.1, the absolute values of these energies are equal to the energies necessary to ionize the electron occupying each orbital. Further discussion will be made in Sect. 8.2.3 below concerning the comparison of theoretically determined energy values and those obtained experimentally.

In Table 8.1 the results for ten-electron molecules other than HF are also listed. The illustrations to the right of the table show schematically the basis functions of significant weights. For example, the first MO of 1σ is depicted by an open and a shaded circle which crudely represent the coefficients for χ_{F1s} and χ_{F2s}.

The coefficient of 0.08 for χ_{H1s} is negligibly small so that the 1σ MO is almost like a sum of the 1s and 2s orbitals of a free F atom. Similarly, in

Table 8.1. Molecular orbitals of HF and H_2O molecules with the minimal basis set. B.J. Ransil (1960), R.M. Pitzer et al. (1967, 1970), W.E. Palke et al. (1966)

HF

MO	AO						ε_i/eV	
	χ_{F1s}	χ_{F2s}	χ_{F2p_x}	χ_{F2p_y}	χ_{F2p_z}	χ_{H1s}		
1σ	0.99	−0.24	0.08	0	0	0.08	−711.2	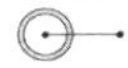 $\rightarrow x$
2σ	0.01	0.93	0.47	0	0	−0.55	−40.2	
3σ	0.00	0.09	0.68	0	0	−0.80	−15.4	
1π	0	0	0	1.00	0	0	−12.6	
1π	0	0	0	0	1.00	0	−12.6	
$4\sigma^*$	−0.00	0.16	0.57	0	0	1.05	+12.9	

H_2O

MO	AO							ε_i/eV	
	χ_{O1s}	χ_{O2s}	χ_{O2p_x}	χ_{O2p_y}	χ_{O2p_z}	$\chi_{\mathrm{H(1)ls}}$	$\chi_{\mathrm{H(2)ls}}$		
$1a_1$	0.99	0.01	0	0	0.00	−0.00	−0.00	−557.9	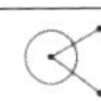 $\rightarrow z$
$2a_1$	−0.22	0.84	0	0	0.13	0.14	0.14	−35.3	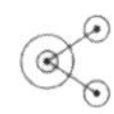
$1b_2$	0	0	0	0.58	0	0.51	−0.51	−17.3	
$3a_1$	0.09	−0.55	0	0	0.77	0.28	0.28	−12.8	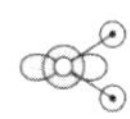
$1b_1$	0	0	1.00	0	0	0	0	−11.5	
$4a_1^*$	−0.11	0.86	0	0	0.70	−0.74	−0.74	+10.9	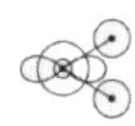

the 3σ MO the major contributions are from χ_{F2p_x} and χ_{H1s}, which overlap each other with the same signs so that the character of the orbital becomes bonding. In contrast, the $4\sigma^*$ MO is anti-bonding and the associated orbital energy becomes positive.

8.2.2 Some Comments on Point Groups

As stated at the beginning of Chap. 8, a systematic discussion of group theory is not covered in this book, but at a minimum, essential explanations related to the group theory will be made here. In the equilibrium nuclear arrangement the molecules of HF, H_2O, NH_3, and CH_4 are linear, equilateral triangular,

Table 8.1. (continued) Molecular orbitals of NH_3 and CH_4 molecules with the minimal basis set. B.J. Ransil (1960), R.M. Pitzer et al. (1967, 1970), W.E. Palke et al. (1966)

NH_3

MO	AO								ε_i/eV	
	χ_{N1s}	χ_{N2s}	χ_{N2p_x}	χ_{N2p_y}	χ_{N2p_z}	$\chi_{H(1)1s}$	$\chi_{H(2)1s}$	$\chi_{H(3)1s}$		
$1a_1$	0.99	0.22	0	0	0.00	−0.00	−0.00	−0.00	−557.9	 → z
$2a_1$	−0.20	0.73	0	0	0.13	0.15	0.15	0.15	−35.3	
$1e_x$	0	0	−0.59	0	0	−0.50	0.25	0.25	−17.3	
$1e_y$	0	0	0	−0.59	0	0	−0.43	0.43	−12.8	
$3a_1$	0.08	−0.45	0	0	0.89	0.12	0.12	0.12	−11.5	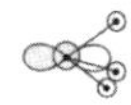
$4a_1^*$	−0.16	1.27	0	0	0.56	−0.71	−0.71	−0.71	+10.9	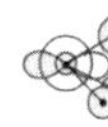

CH_4

MO*	AO									ε_i/eV
	χ_{C1s}	χ_{C2s}	χ_{C2p_x}	χ_{C2p_y}	χ_{C2p_z}	$\chi_{H(1)1s}$	$\chi_{H(2)1s}$	$\chi_{H(3)1s}$	$\chi_{H(4)1s}$	
$1a_1$	0.99	0.02	0	0	0	−0.00	−0.00	−0.00	−0.00	−304.9
$2a_1$	−0.21	0.60	0	0	0	0.18	0.18	0.18	0.18	−25.1
$1t_{2x}$	0	0	0.55	0	0	0.31	0.31	−0.31	−0.31	−14.6
$1t_{2y}$	0	0	0	0.55	0	0.31	−0.31	0.31	−0.31	−14.6
$1t_{2z}$	0	0	0	0	0.55	0.31	−0.31	−0.31	0.31	−14.6

pyramidal, and tetrahedral, respectively. These symmetric molecules are the subject of study of the point group theory. For example, the linear molecule HF belongs to a point group represented by the symbol $C_{\infty v}$.

Before proceeding to a short explanation of the point group theory, a comment will be made on the orbital angular momentum of electrons in linear molecules.

As discussed in Chaps. 3 and 4, the orbital angular momentum of the electron in a spherically symmetric potential field is a physical quantity associated with good quantum numbers. In such a case, the one-electron function to describe the system is a simultaneous eigenfunction of the energy operator $\hat{H}$ and the orbital angular momentum operators $\hat{l}^2$ and $\hat{l}_z$ (cf. Sect. 2.1.12).

However, when the symmetry of the potential field lowers from spherical to axial, the electronic function is no longer an eigenfunction of the operator $\hat{l}^2$. This is because in the axial symmetry the quantum number l to determine the orbital angular momentum of the electron remains no longer constant, in contrast to the situation in Figs. 3.6 and 3.7. However, even in such a case, the electron circulating around the axis remains as an eigenfunction of the operator $\hat{l}_z$. The states with eigenvalues of $m = 0, 1, 2, \ldots$ for $\hat{l}_z$ are customarily designated by $\sigma, \pi, \delta, \ldots$ after the convention to represent the states with eigenvalues of $l = 0, 1, 2, \ldots$ for the orbital angular momentum operator by the letters $\mathrm{s}, \mathrm{p}, \mathrm{d}, \ldots$, respectively. In the states of $\sigma, \pi, \delta, \ldots$ the number of the nodal planes through the molecular axis is $0, 1, 2, \ldots$, respectively. This can be confirmed by looking at the illustrations 1σ through $4\sigma^*$, and 1π for HF in Table 8.1. The Arabic numbers preceding the same Greek letters indicate the order of increasing orbital energy.

Molecules belonging to the $C_{\infty \mathrm{v}}$ point group do not change their atomic positions upon the mirror image operation with regard to any plane including the molecular axis and upon the rotation around the axis by any angle. As is seen from the MO patterns of HF, orbitals of the σ type do not change their sign upon the reflection against the mirror while the orbitals of the π type change the sign. This definite difference with regard to mirror image suggests that the electronic state of linear molecules can be discriminated according to the effect of the symmetry operation, alluding to the fact that in linear molecules the effect of the symmetry operation is closely related to the orbital angular momentum of the molecules.

In molecules other than HF listed in Table 8.1, there is no chance to think about the conservation of the orbital angular momentum any more. Nevertheless, for molecules belonging to a point group such as $C_{2\mathrm{v}}(H_2O)$, $C_{3\mathrm{v}}(NH_3)$, and $T_{\mathrm{d}}(CH_4)$, they can be classified by way of the effects of the symmetry operation of each point group.

As (8.1) and (8.2) indicate, the kinetic energy operator of the electron and the operator to represent the interelectronic repulsion are not altered by symmetry operations so that the total Hamiltonian $\hat{H}$ of the molecule remains unchanged through any symmetry operation.

If we regard symmetry operations to the molecular orbital as a type of quantum mechanical operations, they may be symbolized as $\hat{O}_R$ where the suffix R differentiates each operation from the others. Then, the above statement is equivalent to saying that the two quantum mechanical operators $\hat{H}$ and $\hat{O}_R$ are commutative.

A concrete example will be shown for H_2O. The symmetry operations for the H_2O molecule include;

(1) Doing nothing (the identity operation $\hat{E}$)
(2) Rotation around the z-axis by 180° where the axis bisects the molecule ($\hat{C}_2(z)$)
(3) A mirror-image reflection which contains the bisecting axis and is perpendicular to the molecular plane ($\hat{\sigma}_{\mathrm{v}}(zx)$)
(4) Reflection with regard to the molecular plane ($\hat{\sigma}_{\mathrm{v}}(yz)$).

It should be apparent that these four operations are commutative (e.g., compare the results of $\hat{C}_2(z)\hat{\sigma}_v(zx)$ and $\hat{\sigma}_v(zx)\hat{C}_2(z)$). Consider that these operations are effected on the molecule. The identity operation gives the relation $\hat{O}_E\varphi = \varphi$. As for the remaining three operations, the relation $\hat{O}_R^2\varphi = \varphi$ holds so that $\hat{O}_R\varphi$ is either $+\varphi$ or $-\varphi$. According to the group theory, among the various combinations of the sign of φ, only the four in Table 8.2(a) are meaningful.

Tables such as those in Table 8.2 are called character tables in the group theory. The symbols in Table 8.2(a), A_1 through B_2, have no physical meanings, contrary to the symbols of $\sigma, \pi, \delta, \ldots$ which have a relationship with the orbital angular momentum as stated above for HF, belonging to the point group of $C_{\infty v}$. The letters A and B simply differentiate the symmetric (+) and the anti-symmetric (−) behavior with regard to the operation of $\hat{C}_2(z)$ on molecules such as H_2O. Similarly, the suffixes 1 and 2 are to differentiate the effects of the remaining two operations of $\hat{\sigma}_v(zx)$ and $\hat{\sigma}_v(yz)$.

All that the table shows is that the results of the four operations on an object (function) are classified into the four kinds of A_1 through B_2. These are generally called the symmetry species or the irreducible representations. As mentioned in Sect. 1.2, quantum mechanical operators manifest their significance by giving the result of the operation on any object such as an eigenfunction. The effect of symmetry operations in the group theory is similar to the effect of the quantum mechanical operators in that the irreducible representation tells how the operated object is converted by the set of symmetry operations. The object here is often called the basis. Another explanation would be as follows; in a point group with no degeneracy such as C_{2v}, the characters such as those in Table 8.2 are the eigenvalues of the operators, i.e., the symmetry operations. In the case of C_{2v}, a given function is the simultaneous eigenfunction of four operators (symmetry operations) and the set of the eigenvalues (characters) constitute an irreducible representation. The set can be regarded as expanding four orthonormal basis vectors in the four dimensional vector space (cf. the orthogonality theorem of characters in the group theory*).

Table 8.2. Character tables of C2v and T_d

(a) C_{2v}

	Operation			
Species	$\hat{E}$	$\hat{C}_2(z)$	$\hat{\sigma}_v(zx)$	$\hat{\sigma}_v(yz)$
A_1	1	1	1	1
A_2	1	1	−1	−1
B_1	1	−1	1	−1
B_2	1	−1	−1	1

(b) T_d

	Operation				
Species	$\hat{E}$	$8 \times \hat{C}_3$	$3 \times \hat{C}_2$	$6 \times \hat{S}_4$	$6 \times \hat{\sigma}_d$
A_1	1	1	1	1	1
A_2	1	1	1	−1	−1
E	2	−1	2	0	0
T_1	3	0	−1	1	−1
T_2	3	0	−1	−1	1

In Table 8.2 the irreducible representations are expressed by capital letters while those in Table 8.1 are in small letters. This is in obedience to the convention that symmetry representations for a one-component function, e.g., a one-electron function in multi-electron systems are represented by a small letter.

As for the point groups having some degeneracy, a set of several bases belong to the same irreducible representation. Examples are $1e_x$ and $1e_y$ for NH_3 and $1t_{2x}$ through $1t_{2z}$ for CH_4 in Table 8.1. Any new set obtainable by a linear (orthogonal) transformation of the degenerate set can describe the system as equally as the original set. In the above examples, the numbers in Table 8.1 for NH_3 and CH_4 vary in infinite ways depending upon the choice of the transformation (cf. Fig. 2.3 in Sect. 2.2.6). As the symmetry of a molecule becomes higher, degeneracies occur in physical quantities such as the energy. Correspondingly, the basis of the irreducible representation contains several functions. The point group for atoms dealt with in Chaps. 3 and 4 is the spherical group denoted by K_{h}. As has been shown for atomic systems, the angular momentum also is degenerate as is the energy. This is a reflection of the fact that the point group K_{h} is for the most symmetric object.

8.2.3 Computed Orbital Energy as Compared with Experimental Observation

According to the discussion in the preceding Sect. 8.2.2, MOs constituting the molecular eigenfunction in (8.6) belong to irreducible representations of the point group appropriate to that molecule. As stated in Sect. 8.1.2, these orbitals are called the canonical molecular orbitals with their orbital energies ε_i being related to the ionization energies E_{i} as $\varepsilon_i \cong -E_{\mathrm{i}}$ on account of Koopmans' theorem.

The spectra in Fig. 8.1 show the relation between the ionization energy E_{i} and the relative intensity of ionized electrons, i.e., the emission probability of the electron. Such spectra are called photoelectron spectra.

The photoelectron spectrum is obtained by irradiating the target atom or molecule in the gas phase with photons of a known energy of $h\nu$, and the kinetic energy E_{k} of the released electrons (called photoelectrons) is measured by suitable detectors. By the conservation of energy, the relation $h\nu = E_{\mathrm{k}} + E_{\mathrm{i}}$ holds, from which the sought E_{i} is immediately determined. Then, the orbital energy ε_i is given by the relation $\varepsilon_i \cong -E_{\mathrm{i}}$. The quantity $|\varepsilon_i|$ may be regarded as the binding energy of the released electron when it is in an atom or a molecule.

The photoelectron spectrum can also be regarded as a spectrum showing the energy level of the ionized atom or molecule with the origin of the abscissa being associated with the ground electronic state of the ion.

Inspection of the individual spectra in Fig. 8.1 reveals that the spectral width for Ne is so sharp that it appears to be a line spectrum, whereas the spectra for the molecules are mostly broadened to some extent, showing

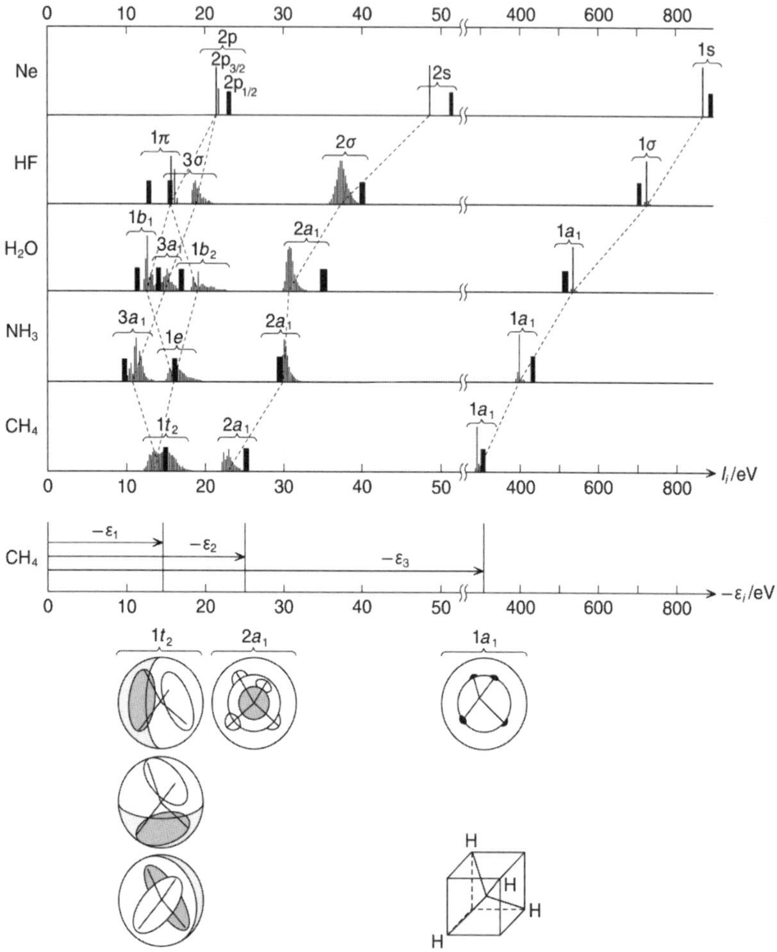

Fig. 8.1. Photoelectron spectra and the corresponding orbital energies of some ten-electron atom and molecules.
Top: The observed photoelectron spectra of Ne through CH4 and the corresponding ionization energies (reproduced from J.W. Rabalais, *Principles of Ultraviolet Photoelectron Spectroscopy*, John Wiley & Sons, 1977. Some of the broken connecting lines in the original book have been rectified). The curly brackets designated by the orbital symbols demonstrate that the calculated orbital energies (ε_i) in bold sticks correspond to the experimental ionization energies in thin sticks. The fair agreement between the calculated and the experimental sticks bears out Koopmans' theorem. The broadening of some experimental sticks manifests the vibrational and rotational structure of the molecular ions HF^+ through CH_4^+ in each electronic state.
Middle: Reproduction of the thick stick spectrum for CH_4. The energies indicated by $-\varepsilon_1$, $-\varepsilon_2$, and $-\varepsilon_3$ correspond, respectively, to the ionization energies from the orbitals $1t_2$, $2a_1$ and $1a_1$.
Bottom: Schematic illustrations of the patterns of the molecular orbitals $1t_2$, $2a_1$ and $1a_1$, when the hydrogen atoms are arranged as shown in the lower right corner

spectral structure due to the vibration and rotation of the ions (cf. Fig. 5.1). As seen from the figure, the ionization energies of the four molecules decrease when the central atoms change as F, O, N, and C. This tendency of the decrease coincides with the tendency of the decrease in the electronegativity derived from various sources of chemical behavior of the four atoms.

As stated in the caption, the bold sticks in Fig. 8.1 show the SCF-MO energies $\varepsilon_i(\cong - E_{\rm i})$ computed by the Hartree-Fock-Roothaan method explained in Sect. 8.2.1 for HF. For the four molecules, see the fourth column of Table 8.1. As for Ne, see Table 4.3 in Chap. 4 (C.F. Fischer, 1977). Despite several deficiencies inherent to the computation which are mentioned at the beginning of Sect. 8.1, Fig. 8.1 demonstrates that the agreement of the computed energies with the observed ionization energies is relatively good.

Since Ne in Fig. 8.1 is a multi-electron atom dealt with in Chap. 4, the spectrum in Fig. 8.1 will be examined in reference to Chap. 4. First, we note that the ionization energies from the 2s and 2p orbitals belonging to the same $n = 2$ subshells differ by as much as 20 to 30 eV. This fact is a reflection of the removal of the accidental degeneracy, which is proper to one-electron atoms as discussed in Sect. 4.2.

Next, we note the splitting of $2p_{3/2}$ and $2p_{1/2}$ (two thin sticks at about 21.5 eV). This is also understood by referring to Hund's rule (3) in Sect. 4.3.4; the electronic structure in the 2p subshell of the ionized Ne is $2p^5$, so that according to the rule, the energy level of $J = 3/2$ for Ne^+ should be lower than that of $J = 1/2$ and the observed doublet results.

These two levels with the multiplicities $2J + 1$ of 4 and 2 are degenerate in the absence of an external magnetic field as discussed in Sect. 4.3.2. The total degeneracy of $4 + 2$ agrees with the multiplicity given by the relation $(2L + 1)(2S + 1)$ with $L = 1$ and $S = 1/2$ for the LS-coupling case when spin-orbit interaction is ignored.

Let us consider the implication of the broken connecting lines in Fig. 8.1. The two lines starting from the two levels of $J = 3/2$ and $1/2$ of Ne^+ are correlated to the two energy states of HF^+ which correspond to the MOs 1π and 3σ of HF in Table 8.1. The four-fold degeneracy of the level of $J = 3/2$ in Ne^+ comprises the four energy states of $J = 3/2$, $M = \pm 3/2$ and $J = 3/2$, $M = \pm 1/2$ while the two-fold degenerate level of $J = 1/2$ consists of the two energy states of $J = 1/2$, $M = \pm 1/2$. Therefore, the energy states characterized by $M = \pm 1/2$ have two origins associated with the levels of $J = 3/2$ and $J = 1/2$. In contrast, the energy states characterized by $M = \pm 3/2$ originate only from the level of $J = 3/2$.

As for the MO 3σ in HF, it consists mainly of the $2p_x$ of the F atom (cf. Table 8.1). The remaining two 2p orbitals, $2p_y$ and $2p_z$, are perpendicular to the molecular axis. They provide a doubly degenerate pair of 1π molecular orbitals of HF.

The comparison between the orbitals of Ne and HF, therefore, indicates that the non-degenerate 3σ MO of HF correlates with the $J = 3/2$, $M = \pm 3/2$

energy states of Ne^+, and that the doubly degenerate 1π MOs of HF correlate with two sets of energy states of Ne^+, i.e., $J = 3/2$, $M = \pm 1/2$ and $J = 1/2$, $M = \pm 1/2$.

In the above correlation between Ne^+ and HF^+, we have referred to the angular momentum of both ions. In the cases of the four molecules in Fig. 8.1, the correlation can be determined automatically by way of the group theoretical manipulation to resolve degenerate irreducible representations of a point group of a higher symmetry to those of a lower symmetry point group. Such a correlation diagram is easily found in textbooks of elementary point group theory or molecular spectroscopy (cf. e.g., G. Herzberg, 1991).

It should be clear by now that photoelectron spectra such as the one in Fig. 8.1 are instructive for understanding the orbital energy and its connection with the molecular symmetry.

8.2.4 The Relationship between Canonical Molecular Orbitals and Localized Orbitals

So far we have mainly discussed how to determine and interpret the eigenfunction Φ, which has been expressed through a single Slater determinant in (8.6). The eigenfunction is considered to be constructed from one-electron functions ϕ, which are approximated by LCAO-type trial functions φ. In the course of this mathematical process, the trial function has been named the canonical molecular orbital. Now we will consider the canonical molecular orbital in relation to such common statements as "the two OH bonds in the H_2O molecule are equivalent", which is familiar to chemists.

Actually, the relationship is easily understood through reference to the following two propositions.

(1) The eigenfunction Φ is approximated by a single Slater determinant.
(2) The value of any determinant remains the same when any row or column, after a multiplication with a common factor, is added to or subtracted from any other row or column.

Let us look at the tangible example presented by the H_2O molecule. The abbreviated symbols for the canonical molecular orbitals in Table 8.1, such as $1a_1$, will be written hereafter as ϕ_{1a_1}. Then, the determinant corresponding to (8.6) becomes:

$$\begin{aligned}\Phi(1,2,\ldots,10) = \|&\phi_{1a_1}(1)\overline{\phi_{1a_1}(2)}\phi_{2a_1}(3)\overline{\phi_{2a_1}(4)}\phi_{1b_2}(5)\overline{\phi_{1b_2}(6)}\\ &\times\phi_{3a_1}(7)\overline{\phi_{3a_1}(8)}\phi_{1b_1}(9)\overline{\phi_{1b_1}(10)}\|\end{aligned} \tag{8.31}$$

According to Fig. 8.1 and Table 8.1, electrons occupying the canonical molecular orbital $\phi_{1a_1} (\equiv 1a_1)$ are associated with a binding energy of about 500 eV. These electrons are localized almost completely in the 1s orbital of the O atom so that they scarcely contribute to chemical bond formation. Such orbitals are called, in general, the inner-shell orbitals. In contrast, electrons in the canonical molecular orbitals $2a_1, 1b_2, 3a_1$, and $1b_1$ involve themselves significantly in bonding. Consequently, these orbitals are called the valence orbitals. We will ignore the inner-shell orbitals and focus on the valence orbitals in this discourse.

According to the second proposition above, the determinant used to represent the eigenfunction in (8.31) can be rewritten as below, e.g., without a change in the value of the determinant.

$$\begin{aligned}\Phi = \Big\| &\phi_{1a_1}\overline{\phi_{1a_1}(2)} \\ &\times \left\{(1/\sqrt{2})(\phi_{2a_1}(3)+\phi_{1b_2}(3))(1/\sqrt{2})\,\overline{(\phi_{2a_1}(4)+\phi_{1b_2}(4))}\right\} \\ &\times \left\{(1/\sqrt{2})(\phi_{2a_1}(5)-\phi_{1b_2}(5))(1/\sqrt{2})\,\overline{(\phi_{2a_1}(6)-\phi_{1b_2}(6))}\right\} \\ &\times \left\{(1/\sqrt{2})(\phi_{3a_1}(7)+\phi_{1b_1}(7))(1/\sqrt{2})\,\overline{(\phi_{3a_1}(8)+\phi_{1b_1}(8))}\right\} \\ &\times \left\{(1/\sqrt{2})(\phi_{3a_1}(9)-\phi_{1b_1}(9))(1/\sqrt{2})\,\overline{(\phi_{3a_1}(10)-\phi_{1b_1}(10))}\right\}\Big\| \end{aligned} \tag{8.32}$$

This determinant signifies that the following four one-electron functions will accommodate electrons 3 through 10.

$$\lambda_1(1) = (\phi_{2a_1}(1)+\phi_{1b_2}(1))/\sqrt{2} \tag{8.33}$$

$$\lambda_2(1) = (\phi_{2a_1}(1)-\phi_{1b_2}(1))/\sqrt{2} \tag{8.34}$$

$$\lambda_3(1) = (\phi_{3a_1}(1)+\phi_{1b_1}(1))/\sqrt{2} \tag{8.35}$$

$$\lambda_4(1) = (\phi_{3a_1}(1)-\phi_{1b_1}(1))/\sqrt{2} \tag{8.36}$$

It is easily confirmed that these are orthogonal in reference to the orthogonality of the original orbitals $\phi_{2a_1}, \phi_{1b_2}, \ldots$. The eigenfunction in (8.32) can be concisely written as:

$$\Phi = \left\|\phi_{1a_1}(1)\overline{\phi_{1a_1}(2)}\lambda_1(3)\overline{\lambda_1(4)}\lambda_2(5)\overline{\lambda_2(6)}\lambda_3(7)\overline{\lambda_3(8)}\lambda_4(9)\overline{\lambda_4(10)}\right\| \tag{8.37}$$

The patterns of the new orbitals λ_1 through λ_4 are roughly sketched as in Fig. 8.2 in reference to the canonical molecular orbital patterns shown to the right of Table 8.1.

Figure 8.2 suggests that the two OH bonds and the two nonbonding electron pairs (or the lone pair electrons) familiar in elementary chemistry may be approximated by suitable linear combinations of the canonical molecular orbitals. Since (8.33) through (8.36) are the simplest combinations, the OH bonds as well as the nonbonding electron pairs are only crudely described as in Fig. 8.2.

Nevertheless, it is important to recognize that linear combinations of the canonical molecular orbitals can yield localized orbitals or equivalent orbitals such as the above pair of λ_1 and λ_2, and the pair of λ_3 and λ_4. The detailed procedure to determine the coefficients of the linear combinations, which reproduce well the real OH bonds and the nonbonding electron pairs, will not

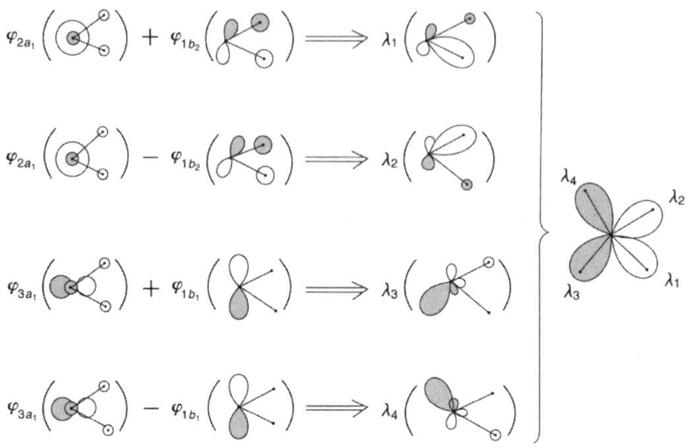

Fig. 8.2. Schematic diagram to show that the localized orbitals λ_1 through λ_4 are formed by linear combinations of the canonical molecular orbitals

be given here, however the general technique is to find a suitable orthogonal transformation from a set of n canonical molecular orbitals to n orthogonal localized orbitals whose orbital energies are maximized. Concrete criteria for choosing such a transformation have been discussed in the literature (J.M. Foster, S. Boys (1960); C. Edmiston, K. Ruedenberg (1963)).

As an example, the result of the localized orbitals obtained for the H_2O molecule with the use of the minimal basis set will be given in (8.38) through (8.42), which are reproduced from the textbook cited previously (I.N. Levine, 2000) with minor adaptations.

The symbols $i(\mathrm{O})$, $b(\mathrm{OH})$, and $l(\mathrm{O})$ in these equations represent the inner-shell orbital of the O atom, the OH bonding orbital, and the lone pair orbital of the O atom, respectively. The underlined terms are major terms, by which the essential features of the localized orbitals are characterized.

$$\begin{aligned} i(\mathrm{O}) = & -0.007\chi_{\mathrm{H(1)1s}} - 0.007\chi_{\mathrm{H(2)1s}} + \underline{1.018\chi_{\mathrm{O1s}}} - 0.123\chi_{\mathrm{O2s}} \\ & +0.03\chi_{\mathrm{O2p}_z} \end{aligned} \tag{8.38}$$

$$\begin{aligned} b(\mathrm{OH}_1) = & \ \underline{0.50\chi_{\mathrm{H(1)1s}}} - 0.10\chi_{\mathrm{H(2)1s}} - 0.039\chi_{\mathrm{O1s}} + \underline{0.257\chi_{\mathrm{O2s}}} \\ & \underline{+0.407\chi_{\mathrm{O2p}_z}} + \underline{0.441\chi_{\mathrm{O2p}_y}} \end{aligned} \tag{8.39}$$

$$\begin{aligned} b(\mathrm{OH}_2) = & -0.10\chi_{\mathrm{H(1)1s}} + \underline{0.50\chi_{\mathrm{H(1)1s}}} - 0.039\chi_{\mathrm{O1s}} + \underline{0.257\chi_{\mathrm{O2s}}} \\ & \underline{+0.407\chi_{\mathrm{O2p}_z}} - \underline{0.441\chi_{\mathrm{O2p}_y}} \end{aligned} \tag{8.40}$$

$$\begin{aligned} l_1(\mathrm{O}) = & -0.09\chi_{\mathrm{H(1)1s}} - 0.09\chi_{\mathrm{H(2)ls}} - 0.059\chi_{\mathrm{O1s}} + \underline{0.647\chi_{\mathrm{O2s}}} \\ & \underline{-0.39\chi_{\mathrm{O2p}_z}} - \underline{0.71\chi_{\mathrm{O2p}_x}} \end{aligned} \tag{8.41}$$

$$l_2(\mathrm{O}) = -0.09\chi_{\mathrm{H(1)1s}} - 0.09\chi_{\mathrm{H(2)1s}} - 0.059\chi_{\mathrm{O1s}} + \underline{0.647\chi_{\mathrm{O2s}}} \\ -\underline{0.39\chi_{\mathrm{O2p}_z}} + \underline{0.71\chi_{\mathrm{O2p}_x}} \tag{8.42}$$

With the use of the above localized orbitals the eigenfunction of the whole H_2O system can be written as below.

$$\Phi = \left\| i(\mathrm{O})\overline{i(\mathrm{O})}b(\mathrm{OH}_1)\overline{b(\mathrm{OH}_1)}b(\mathrm{OH}_2)\overline{b(\mathrm{OH}_2)}l_1(\mathrm{O})\overline{l_1(\mathrm{O})}l_2(\mathrm{O})\overline{l_2(\mathrm{O})} \right\| \tag{8.43}$$

Equation (8.43) is, of course, equivalent to (8.31) according to the second proposition mentioned at the beginning of this section.

Chemists have long used dashes to represent a chemical bond between adjacent atoms. Therefore, the localized orbitals presented here may be more familiar than the delocalized molecular orbitals in describing chemical bonds. On the other hand, the delocalized one-electron orbital is better suited to problems involving the electronic excitation of molecules as a whole. Also, the molecular orbital is associated with a definite orbital energy which can be measured approximately by photoelectron spectroscopic measurements as has been discussed in Sect. 8.2.3.

The coefficients to determine a localized orbital can be chosen freely within the orthogonal transformation so that one can prepare localized orbitals most suitable for chemical bond problems. However, a definite value of energy cannot be associated with localized orbitals contrary to the canonical molecular orbital. This is because the localized orbital is a superposition of several canonical orbitals with different orbital energies. It may be said that this situation is a reflection of the Heisenberg uncertainty principle; electrons in localized orbitals are confined to some limited locale in a molecule, which makes the variance $\Delta \boldsymbol{r}$ of the position of the electron smaller compared with the variance in canonical molecular orbitals delocalized over the whole molecule. Accordingly, the variance of momentum of the electron $\Delta \boldsymbol{p}$ in localized orbitals becomes larger, resulting in a larger variance of the kinetic energy of the electron (cf. Sect. 6.2 for a related discussion). In other words, any attempt to interpret photoelectron spectra in terms of localized orbitals is meaningless.

8.2.5 A Critical Review of the Hybridized Orbital of Pauling

A brief comment was made on the hybridized orbital in Sect. 4.3.5 in connection with the valence state of multi-electron atoms. The concept of the hybridized orbital is popular in elementary textbooks, but it seems worthwhile to add some critical commentary in connection with the delocalized canonical molecular orbital. First, procedures to construct a hybridized orbital will be explained, and then, relationships between a hybridized orbital and a molecular orbital will be reviewed.

As a substantive example, we will consider the CH_4 molecule. Denoting the hybridized orbitals and the atomic orbitals with ϕ and χ, respectively, and requiring that the two equivalent hybridized orbitals ϕ_1 and ϕ_2 are orthonormalized, we obtain the following (8.44) through (8.47).

$$\phi_1 = a\chi_{\mathrm{s}} + b\chi_{\mathrm{p}_x} + c\chi_{\mathrm{p}_y} + d\chi_{\mathrm{p}_z} \tag{8.44}$$

$$\phi_2 = a'\chi_{\mathrm{s}} + b'\chi_{\mathrm{p}_x} + c'\chi_{\mathrm{p}_y} + d'\chi_{\mathrm{p}_z} \tag{8.45}$$

$$a^2 + b^2 + c^2 + d^2 = a'^2 + b'^2 + c'^2 + d'^2 = 1 \tag{8.46}$$

$$aa' + bb' + cc' + dd' = 0 \tag{8.47}$$

Portions of the sums in (8.44) and (8.45), i.e., $b\chi_{\mathrm{p}_x} + c\chi_{\mathrm{p}_y} + d\chi_{\mathrm{p}_z}$ and $b'\chi_{\mathrm{p}_x} + c'\chi_{\mathrm{p}_y} + d'\chi_{\mathrm{p}_z}$, can be regarded as two vectors in the three dimensional space (cf. the first three members in Table 3.2). From the definition of the dot products (or the inner products) of vectors, the following relationship is obtained, where the angle θ is between the two vectors contained in ϕ_1 and ϕ_2 in (8.44) and (8.45), respectively.

$$\frac{bb' + cc' + dd'}{\sqrt{b^2 + c^2 + d^2}\sqrt{b'^2 + c'^2 + d'^2}} = \cos\theta \tag{8.48}$$

Since ϕ_1 and ϕ_2 are assumed to be equivalent, the coefficients a through d and a' through d', which are regarded as components of the vectors, satisfy the following relationship.

$$(b^2 + c^2 + d^2)/a^2 = (b'^2 + c'^2 + d'^2)/a'^2 \equiv n \tag{8.49}$$

From (8.46) through (8.49), the following four relations are immediately obtained:

$$a^2 = a'^2 = 1/(n+1) \tag{8.50}$$

$$b^2 + c^2 + d^2 = b'^2 + c'^2 + d'^2 = n/(n+1) \tag{8.51}$$

$$bb' + cc' + dd' = -1/(n+1) \tag{8.52}$$

$$1 + n\cos\theta = 0 \tag{8.53}$$

When the four H atoms occupy the four vertexes of a tetrahedron, $\cos\theta$ is determined to be $-1/3$, from which n is fixed to be 3. In order to determine all the coefficients a through d and a' through d', it is necessary to specify the positions of the H atoms explicitly.

In the case of Fig. 8.3(a), the relations $b = -b'$, $c = -c'$, $d = d'$ and $b = c$ hold for the two CH bonds, CH(1) and CH(2). The substitution of these relations to (8.49) through (8.52) yields the results $b = c = d = 1/2$, $b' = c' = -1/2$ and $d' = 1/2$.

The above results are equivalent to the expressions in (4.9) and (4.10) in Sect. 4.3.5. Similar treatment of CH(3) and CH(4) in Fig. 8.3(a) provides the

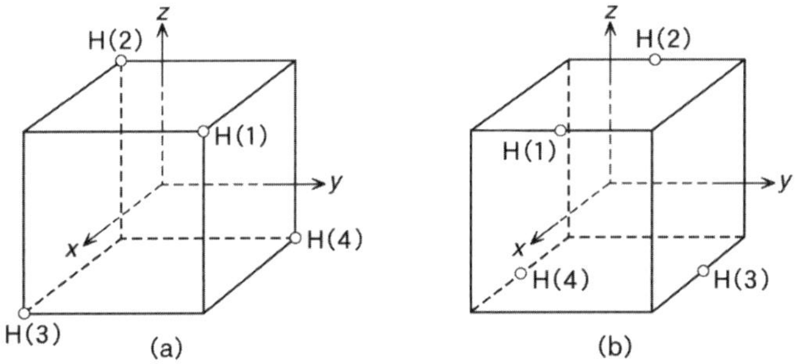

Fig. 8.3. Two arrangements of the four H atoms in CH_4 in the Cartesian coordinate system

remaining two (4.11) and (4.12). If the arrangement of Fig. 8.3(b) is preferred to (a), the equations can be rearranged to provide the following forms instead.

$$\phi_1 = \frac{1}{2}\chi_{\mathrm{s}} + \frac{1}{\sqrt{2}}\chi_{\mathrm{p}_x} + \frac{1}{2}\chi_{\mathrm{p}_z} \tag{8.54}$$

$$\phi_2 = \frac{1}{2}\chi_{\mathrm{s}} - \frac{1}{\sqrt{2}}\chi_{\mathrm{p}_x} + \frac{1}{2}\chi_{\mathrm{p}_z} \tag{8.55}$$

$$\phi_3 = \frac{1}{2}\chi_{\mathrm{s}} + \frac{1}{\sqrt{2}}\chi_{\mathrm{p}_y} - \frac{1}{2}\chi_{\mathrm{p}_z} \tag{8.56}$$

$$\phi_4 = \frac{1}{2}\chi_{\mathrm{s}} - \frac{1}{\sqrt{2}}\chi_{\mathrm{p}_y} - \frac{1}{2}\chi_{\mathrm{p}_z} \tag{8.57}$$

In CH_4, the tetrahedron centered on the C atom can be rotated freely, keeping the relative arrangement of the four H atoms fixed. Thus, the arrangements in Fig. 8.3 are just two cases out of an infinite number of choices. Of course, in order to obtain the linearly independent hybridized orbitals ϕ_1 through ϕ_4, it is necessary to employ the four orthogonal basis functions of s and p_x through p_z for any arrangement of the nuclei (cf. Sects. 2.1.13 and 2.1.14 for the linear independence).

In the discussion of the atomic valence state in Sect. 4.3.5, it was stated that the C atom supplies four open shell electrons with their spin states randomized by a virtual process called 'the promotion'. These four electrons are now identified with the electrons occupying the four linearly independent hybridized orbitals discussed here. When each one of these electrons from the C atom is combined with the 1s electron of each H atom belonging to the bonds CH(1) through CH(4), the stabilization energy will be gained mainly through the exchange interaction as discussed for the H_2 molecule in Chap. 7. This is the explanation of the formation of the stable CH_4 molecule originally proposed by Pauling (1931).

The concept of the hybridized orbital is powerful in its use in understanding the directionality of chemical bonds. In this sense the work of Pauling is eternal. However, it is also true that there is no necessity to endow a special physical meaning to the hybridized orbital; in order to clarify this statement I will examine the canonical molecular orbitals shown roughly in Table 8.1 and Fig. 8.1 again. Although some terminology from the point group theory will be used, it is hoped that a basic understanding of the relationship between the hybridized orbital and the molecular orbital can be obtained by a reader who may not be familiar with elementary point group theory.

The CH_4 molecule belongs to the point group T_d, the irreducible representations of which are classified as in Table 8.2(b). In the table the first row represents the symmetry operations. For example, the notation of $8 \times \hat{C}_3$ means that there are 8 three-fold symmetry axes in the group. In all, there are 24 symmetry operations in the group T_d. The molecular orbitals of CH_4 obtained by the SCF procedure with the minimal STO basis set are listed in Table 8.1. The table shows that the molecular orbitals $1a_1$ and $2a_1$ consist mainly of the 1s and 2s orbitals of the C atom. These molecular orbitals belong to the symmetry species of A_1 in Table 8.2(b). In a similar manner, the molecular orbitals $1t_{2x}$, $1t_{2y}$, and $1t_{2z}$ in Table 8.1, consisting mainly of the orbitals $2p_x$, $2p_y$, and $2p_z$ in the C atom, are classified as belonging to the symmetry species T_2 in Table 8.2(b).

At this point, we will look at the terms involving the H atomic orbitals in the molecular orbitals of CH_4 in Table 8.1 and prepare the following artificial linear combinations of the H atomic orbitals.

$$\chi_{a_1} = \chi_{H(1)1s} + \chi_{H(2)1s} + \chi_{H(3)1s} + \chi_{H(4)1s} \tag{8.58}$$
$$\chi_{t_{2x}} = \chi_{H(1)1s} + \chi_{H(2)1s} - \chi_{H(3)1s} - \chi_{H(4)1s} \tag{8.59}$$
$$\chi_{t_{2y}} = \chi_{H(1)1s} - \chi_{H(2)1s} + \chi_{H(3)1s} - \chi_{H(4)1s} \tag{8.60}$$
$$\chi_{t_{2z}} = \chi_{H(1)1s} - \chi_{H(2)1s} - \chi_{H(3)1s} + \chi_{H(4)1s} \tag{8.61}$$

With the use of these linear combinations for the four H atoms, the canonical molecular orbitals of the whole CH_4 molecule in Table 8.1 can be expressed as follows, where the coefficients are given to four decimal places, in contrast to the truncated two places in Table 8.1.

$$\varphi_{1a_1} = 0.9947\chi_{C1s} + 0.0256\chi_{C2s} - 0.0047\chi_{a_1} \tag{8.62}$$
$$\varphi_{2a_1} = -0.2158\chi_{C1s} + 0.6037\chi_{C2s} + 0.1865\chi_{a_1} \tag{8.63}$$
$$\varphi_{1t_{2x}} = 0.5539\chi_{C2p_x} + 0.3178\chi_{t_{2x}} \tag{8.64}$$
$$\varphi_{1t_{2y}} = 0.5539\chi_{C2p_y} + 0.3178\chi_{t_{2y}} \tag{8.65}$$
$$\varphi_{1t_{2z}} = 0.5539\chi_{C2p_z} + 0.3178\chi_{t_{2z}} \tag{8.66}$$

Now we are in a position to see the relationship between the above canonical molecular orbitals and the localized orbitals as well as the hybridized orbitals. As shown in Sect. 8.2.4 for the H_2O molecule, the eigenfunction Φ of the whole CH_4 system can be expressed in terms of either the canonical molecular orbitals or the localized orbitals, as long as as both are related by some orthogonal transformation (cf. (8.31) and (8.43)). Let us consider the expression in terms of the localized orbitals. We view the molecule as being constructed by the four CH localized orbitals, each of which accommodates two electrons with their spins in an anti-parallel pairing. Thus, out of the total ten electrons, eight are in the four localized orbitals. The remaining two electrons are essentially associated with the inner-shell orbital, i.e., the 1s orbital of the C atom. These electrons are supposed to contribute insignificantly to the bond formation between C and H. The result of the localization procedure by Edmiston-Ruedenberg (1963) turns out to be as follows.

The inner-shell orbital $\varphi(i.s.)$ (which corresponds to $i(\mathrm{O})$ in (8.38) for H_2O) is given by (8.67),

$$\varphi(i.s.) = 1.0125\chi_{\mathrm{C1s}} - 0.0413\chi_{\mathrm{C2s}} - 0.0253\chi_{a_1} . \tag{8.67}$$

As for the CH localized orbitals, one of the four equivalent orbitals, CH(1), is found to be,

$$\begin{aligned}\varphi(\mathrm{CH}(1)) = -0.0522\chi_{\mathrm{C1s}} &+ \underline{0.3014\chi_{\mathrm{C2s}} + 0.2769(\chi_{\mathrm{C2p}_x} + \chi_{\mathrm{C2p}_y} + \chi_{\mathrm{C2p}_z})} \\ &\underline{\underline{+0.5691\chi_{\mathrm{H(1)1s}}}} - 0.0665(\chi_{\mathrm{H(2)1s}} + \chi_{\mathrm{H(3)1s}} + \chi_{\mathrm{H(4)1s}}) .\end{aligned} \tag{8.68}$$

The remaining three orbitals, CH(2) through CH(4), are similar to that in (8.68), except that the signs of the coefficients of χ_{C2p_x}, χ_{C2p_y}, and χ_{C2p_z} change in the same way as those in (4.10) through (4.12). With the use of these five orbitals, the eigenfunction of the whole CH_4 system Φ can be described similarly to the Slater determinant for the H_2O molecule in (8.43).

What is interesting here is that the coefficients of the singly underlined second and third terms of (8.68), i.e., 0.30 vs. 0.28, are approximately equal. Therefore, the sum of these two terms may be regarded as similar to one of the sp^3 hybridized orbitals, $\phi_1 = (\chi_{2\mathrm{s}} + \chi_{2\mathrm{p}_x} + \chi_{2\mathrm{p}_y} + \chi_{2\mathrm{p}_z})/2$ in (4.9).

Since the first and fifth terms of (8.68) are negligibly small, the localized orbital $\varphi(\mathrm{CH}(1))$ may be regarded as essentially describing the bonding between the sp^3 hybridized orbital and the 1s orbital of H atom 1 shown by the double underline in (8.68). Therefore, the formation of CH bonds in the CH_4 molecule can be explained by canonical molecular orbitals as well as by suitably transformed localized orbitals.

In conclusion, the molecular orbital approach supports the intuitive approach by Pauling on the one hand, and on the other hand, the approach discloses that the concept of hybridization is not an absolute necessity for the understanding of chemical bonding.

8.3 Critical Comments on the Classification of Chemical Bonds

In high school chemistry textbooks, terminology such as the ionic bond, the metallic bond, and the coordinate bond appear along with the covalent bond discussed in Sect. 7.3. These different names reflect the effort of people in the past who tried to understand chemical bonding on the basis of the phenomenological features of the bonds. Now that quantum chemistry teaches us universal principles which govern chemical bonding, it may be instructive to review critically these traditional appelations which originated from empiricism.

8.3.1 Ionic Bond: NaCl, for Example

Even before the advent of quantum mechanics, the concept of ionic bonding was widely accepted as a mode of chemical bonding thanks to the easily understood concept of Coulombic attraction between positive and negative ions. In this section we will review the ionic bond from the quantum mechanical viewpoint, using NaCl as a representative example.

Assume that the two ions, Na^+ and Cl^-, are separated from each other initially. They will approach each other by virtue of the Coulombic attraction. However, when the overlap of the electron clouds of both ions becomes appreciable, the repulsion due to the Pauli exclusion principle prevents further approaching. Therefore, the two ions will find an equilibrium distance determined by the counterbalancing effects, at which NaCl stabilizes as the ionic pair, Na^+Cl^-.

However, the story is not that simple. Even in a qualitative consideration one should anticipate additional stabilization due to the interaction between one ion and the induced electric multipoles on the other ion. In addition, one can never neglect the stabilization due to the dispersion force* even though it may be weak.

In this section we will examine the bonding of NaCl in its ground electronic state in terms of the VB theory discussed in Chap. 7 for the H_2 molecule. In the simplest form of the latter theory, the bond forming electrons 1 and 2 are identified with one 3s electron in $Na(1s^2 2s^2 2p^6 3s)$ and one of the 3p electrons in $Cl(1s^2 2s^2 2p^6 3s^2 3p^5)$. The bonding is, then, approximated by the following two-electron Heitler-London type equation, which is similar to (7.22) for the H_2 molecule.

$$\begin{aligned}\Phi_{\mathrm{cov}} &= \frac{1}{\sqrt{2(1+S^2)}}\left[\chi_{\mathrm{Na3s}}(1)\chi_{\mathrm{Cl3p}_x}(2) + \chi_{\mathrm{Na3s}}(2)\chi_{\mathrm{Cl3p}_x}(1)\right] \\ &\quad \times \frac{1}{\sqrt{2}}[\alpha(1)\beta(2) - \alpha(2)\beta(1)] \end{aligned} \tag{8.69}$$

Among the 3p AOs of the Cl atom only the $3p_x$ parallel to the molecular axis x participates effectively in the bond formation. The suffix cov implies that the electrons 1 and 2 are used to form the covalent bond.

The above equation can be rewritten in terms of two Slater determinants just as in (7.24) for the H_2 molecule.

$$\Phi_{\mathrm{cov}} = \frac{1}{\sqrt{2(1+S^2)}}\left[\|\chi_{\mathrm{Na3s}}(1)\overline{\chi_{\mathrm{Cl3p}_x}(2)}\| - \|\overline{\chi_{\mathrm{Na3s}}(1)}\chi_{\mathrm{Cl3p}_x}(2)\|\right] \tag{8.70}$$

In the discussion of the ground state H_2 molecule in Sect. 7.2, it was concluded that the covalent bond expressed by (7.22) or (7.24) is the major contribution to the bond formation. Furthermore, it was concluded that when the ground state H_2 molecule is analyzed in the framework of the MO theory, the ionic contributions represented by the first and the fourth terms in (7.29) are overestimated.

In contrast to the H_2 molecule, however, the predominance of the ionic character of the NaCl molecule is obvious because the Cl atom has a high electronegativity and will attract electrons from the Na atom which has a low ionization energy. Such a biasing of the electron will be approximated by the following equation.

$$\Phi_{\rm ion} = \|\chi_{\rm Cl3p_x}(1)\overline{\chi_{\rm Cl3p_x}(2)}\| \tag{8.71}$$

Taking into account these qualitative considerations the eigenfunction of NaCl in the ground electronic state is expected to be described as follows.

$$\Phi \cong c_1\Phi_{\rm cov} + c_2\Phi_{\rm ion} \tag{8.72}$$

The coefficients for this linear combination can be estimated, as usual, by Ritz's calculus of variation, which will obviously predict the relation $|c_2| \gg |c_1|$ at the equilibrium internuclear distance.

At first glance, this relationship at the equilibrium distance $R_{\rm e}$ might seem to hold irrespective of the internuclear distance so that one might obtain the pair of Na^+ and Cl^- at an infinite separation of the two components. In fact, however, the molecule is known separate into a pair of neutral Na and Cl atoms as shown in Fig. 8.4.

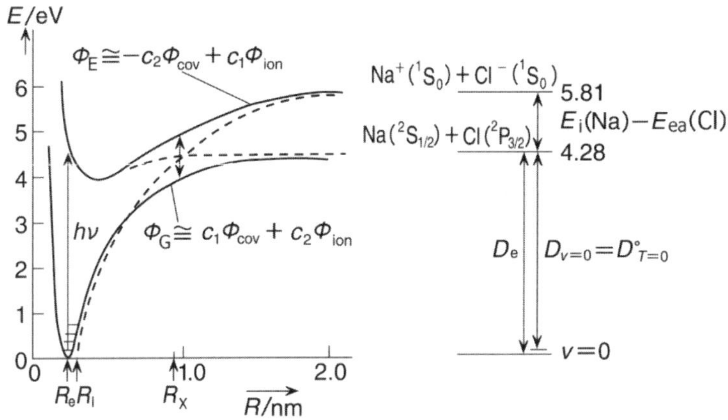

Fig. 8.4. The potential energy curves for the ground electronic state ($\Phi_{\rm G}$) and the first excited electronic state ($\Phi_{\rm E}$) of the NaCl molecule as compared with the energies of the dissociation limits

The lower solid curve in Fig. 8.4 shows that the energy at the dissociation limit of Na+Cl is 4.28 eV, which will be explained shortly in reference to thermodynamic data.

The ionization energy $E_{\mathrm{i}}(\mathrm{Na})$ of the Na atom and the electron affinity $E_{\mathrm{ea}}(\mathrm{Cl})$ of the Cl atom both in their ground electronic state are known from spectroscopic measurements as 5.139 eV and 3.613 eV, respectively. Therefore, the energy at the ionic dissociation limit becomes 5.81 eV as shown in the figure. The electron affinity is defined as the minimum required energy to attach one electron to an atom or a molecule A, resulting in the reaction $\mathrm{A} + \mathrm{e}^- \rightarrow \mathrm{A}^-$. The rather vague terminology of 'affinity' could have been replaced with a more appropriate term such as 'electron attachment energy'. If the heat of reaction ΔH of the above attachment is negative, i.e., if the reaction is exothermic, the positive energy $-\Delta H$ is defined as the electron affinity.

In the case of such an exothermic reaction, the product A^- is thermodynamically stable. The electron affinity of A is, then, determined by knowing the minimum photon energy $h\nu$ in the photoionization process of $\mathrm{A}^- h\nu \rightarrow \mathrm{A}+\mathrm{e}^-$. If, on the other hand, the electron attachment $\mathrm{A}+\mathrm{e}^- \rightarrow \mathrm{A}^-$ is endothermic, one way to estimate the electron affinity (a negative quantity) is to resort to reliable quantum chemical computations of the total energies of A and A^-, the difference of which is identifiable as the electron affinity.

In the present case of NaCl, reference to the above cited $E_{\mathrm{i}}(\mathrm{Na})$ and $E_{\mathrm{ea}}(\mathrm{Cl})$ reveals that the ionic dissociation limit of $\mathrm{Na}^+ + \mathrm{Cl}^-$ is unstable relative to the neutral dissociation limit of Na + Cl by the amount $E_{\mathrm{i}}(\mathrm{Na}) - E_{\mathrm{ea}}(\mathrm{Cl}) = 1.526\,\mathrm{eV}$, which is shown in Fig. 8.4.

We now explain why the dissociation energy for the neural fragments denoted by D_{e} in Fig. 8.4, becomes 4.28 eV; the dissociation energy of a diatomic molecule AB in the standard state (1 atm, 298.15 K) denoted by $D^{\circ}_{298}(\mathrm{A}-\mathrm{B})$ is defined as in (8.73) below in terms of the standard enthalpy (heat) of formation $\Delta_{\mathrm{f}} H^{\circ}$ of the three chemical species A, B, and AB involved in the reaction $\mathrm{AB} \rightarrow \mathrm{A} + \mathrm{B}$.

$$D^{\circ}_{298}(\mathrm{A}-\mathrm{B}) = \Delta_{\mathrm{f}} H^{\circ}(\mathrm{A}) + \Delta_{\mathrm{f}} H^{\circ}(\mathrm{B}) - \Delta_{\mathrm{f}} H^{\circ}(\mathrm{AB}) \tag{8.73}$$

In (8.73) the superscript $\circ$ indicates that the pressure of the standard state is chosen as 1 atm ($= 1.01325 \times 10^5$ Pa).

When AB is NaCl, the values of $\Delta_{\mathrm{f}} H^{\circ}(\mathrm{Na})$, $\Delta_{\mathrm{f}} H^{\circ}(\mathrm{Cl})$, and $\Delta_{\mathrm{f}} H^{\circ}(\mathrm{NaCl})$ necessary for the determination of $D^{\circ}_{298}(\mathrm{Na}-\mathrm{Cl})$ are given as the enthalpic changes in the reactions of $\mathrm{Na(s)} \rightarrow \mathrm{Na(g)}$, $(1/2)\mathrm{Cl_2(g)} \rightarrow \mathrm{Cl(g)}$, and $\mathrm{Na(g)} + (1/2)\mathrm{Cl_2(g)} \rightarrow \mathrm{NaCl(g)}$, respectively (s; solid, g; gas). These thermochemical values are available with high reliability, from which the dissociation energy is obtained as $D^{\circ}_{298}(\mathrm{Na}-\mathrm{Cl}) = 412.1\,\mathrm{kJ\,mol^{-1}} = 4.27\,\mathrm{eV}$. As for the value at 0 K, not at 298.15 K, the sought for energy for $D^{\circ}_{T=0} \equiv D^{\circ}_0$ can be determined by the relationship $D^{\circ}_{298} = D^{\circ}_0 + (3/2)RT$, where R stands for the gas constant. The energy of $(3/2)RT = 0.038\,\mathrm{eV}$ corresponds to the kinetic energy at the center of gravity of the NaCl molecule, separate from the internal electronic, vibrational, and rotational energies. Thus, the energy D_0° turns out to be $D_0^\circ = 408.4\,\mathrm{kJ\,mol^{-1}} = 4.23\,\mathrm{eV}$. Since the NaCl molecule at $T = 0\,\mathrm{K}$ is exclusively in the ground vibrational state $v = 0$, the energy $D_{T=0}^\circ = D_0^\circ$ corresponds to the energy difference between the zero-point vibrational energy state (cf. Fig. 5.1) and the neutral dissociation limit, i.e., $D_{v=0}^\circ = D_{T=0}^\circ$ in Fig. 8.4. As for the zero-point energy, it is known spectroscopically to be $366\,\mathrm{cm^{-1}} = 0.045\,\mathrm{eV}$. Collecting all these values from the literature, the energy indicated by D_e in Fig. 8.4 is finally estimated as $D_e = (4.23 + 0.045)\,\mathrm{eV} = 4.28\,\mathrm{eV}$, as stated above in advance.

In Fig. 8.4, the energy necessary to excite the NaCl molecule to the excited electronic state Φ_E in the vicinity of the equilibrium internuclear distance should correspond to the photon absorption energy by the gaseous NaCl molecule. As a matter of fact, the observed absorption spectrum shows a broad absorption band starting from about 4.2 eV. The energy difference between the two states, Φ_G and Φ_E, shown by the notation $h\nu$ in Fig. 8.4, is in good agreement with this photon energy.

Let us consider, from a different viewpoint, why the NaCl molecule with a dominant ionic character in its ground electronic state, is not associated with the ionic dissociation limit of $Na^+ + Cl^-$. When the two infinitely separated ions start approaching each other, the potential energy of the ionic pair under the Coulombic attraction decreases in a hyperbolic manner according to the R dependence of $D_e + \{E_i(\mathrm{Na}) - E_{ea}(\mathrm{Cl})\} - e^2/(4\pi\varepsilon_0 R)$. The distance R_I in Fig. 8.4 corresponds to the intersection of this hyperbola with the abscissa. With the known values of D_e and $E_i(\mathrm{Na}) - E_{ea}(\mathrm{Cl})$, this distance is determined to be $R_I = 0.248\,\mathrm{nm}$, which is fairly close to the experimentally determined equilibrium internuclear distance of $R_e = 0.236\,\mathrm{nm}$. The closeness of the two distances is consistent with the ionic character of the NaCl in the ground state.

On the other hand, if the neutral Na and Cl atoms in their ground state approach each other from an infinite separation, the character of covalent bond will manifest itself through the overlap of the electronic clouds of the two atoms. The asymptote, represented by the broken line converging to the Na+Cl limit in Fig. 8.4, represents an approximate solution for the covalent bonding approach. The curve is seen to be almost horizontal until the two atoms approach each other to a distance of about 1 nm. This curve intersects with the hyperbolic curve for the ionic approach at a distance designated by R_X, which is much greater than the equilibrium internuclear distance R_e. At a distance of R_X, the covalent state represented by (8.70) and the ionic state represented by (8.71) are energetically degenerate. Whenever two states are of the same symmetry (the meaning of which will be explained below), such as those represented by the two broken curves here, they do not actually intersect but avoid each other as shown by the two solid curves designated by Φ_G and Φ_E in Fig. 8.4. In the two states represented by the solid curves, the NaCl molecule has both covalent and ionic character to a similar extent at the distance of R_X. At distances other than R_X also, both characters coexist depending on the magnitude of the coefficients c_1 and c_2 in the linear combination of (8.72).

The non-intersection of the two broken curves stated here is associated with the non-crossing rule. The meaning of the 'same symmetry' mentioned above is that the molecular axis components of the orbital angular momentum and the electron spin of NaCl and Na^+Cl^-, viewed as individual states of the same diatomic molecule, coincide. Although detailed explanations on this subject will not be pursued, a thorough analysis on the correlation between the orbital angular momentum component along the molecular axis of diatomic molecules and the orbital angular momentum of the constituent two atoms is made in the literature (E. Wigner, E.E. Witmer (1928); G. Herzberg (1989)).

In the above case of NaCl and Na^+Cl^- the axis components of the orbital angular momentum, whose quantum number is designated by Λ, have the same value, namely $\Lambda = 0$. As for the electron spin, the spin multiplicity of both NaCl and Na^+Cl^- are 1. These states of NaCl and Na^+Cl^-, with $\Lambda = 0$ and the spin multiplicity of 1, are said to have the same symmetry and are expressed compactly as $^1\Sigma$ states, analogous with the expression ^1S for multi-electron atoms.

In connection with the broken hyperbolic potential curve in Fig. 8.4, a related topic bears mentioning. Suppose that the almost ionic NaCl molecule at the equilibrium internuclear distance is rapidly separated. The two ions, Na^+ and Cl^-, are supposed to reach the ionic dissociation limit $Na^+ + Cl^-$ according to the broken hyperbolic curve. For this imaginary process to proceed it must be assumed that the necessary energy is freely supplied. Such a process is called a diabatic process in general. Suppose, conversely, the extreme case of a very slow separation of the two ions. Under such a condition, electrons follow the slow motion of the ions instantly so that the separation proceeds along the lower solid curve of Fig. 8.4. That is to say, the electron on the Cl^- ion may easily jump to the Na^+ ion to switch from the $Na^+ + Cl^-$ curve to the Na+Cl curve. The energy necessary for this change is equal to 4.28 eV, relative to the minimum of the potential curve for the Φ_G state of Fig. 8.4. Since no excess energy (heat) flows to the system, this slow separation is called an adiabatic process.

So far it has been assumed that the two states Φ_G and Φ_E represented by the solid curves in Fig. 8.4 are independent eigenstates with no interaction between them. However, this assumption is based on the ignorance of all interactions other than the interaction due to the simple electrostatic potentials appearing in (8.2) in Sect. 8.1(a). In fact, there are various other interactions, which will not be discussed explicitly here, such that the excited state Φ_E has a finite probability of relaxing to the ground state Φ_G. Such a process is called an nonadiabatic process.

A.H. Zewail earned the Nobel prize in chemistry in 1999 on account of his experimental work related to the above processes (A.H. Zewail et al., 1988); he irradiated gaseous NaI molecules with short laser pulses of some 100 femto (10^{-15}) seconds to excite the molecules from the Φ_G to Φ_E the state just like the vertical arrow in Fig. 8.4 shows. The vibrational wave packet of the excited NaI molecules oscillated in the adiabatic well of the Φ_E state with an average oscillation period of 1.25 pico (10^{-12}) seconds and eventually crossed over to the ground state Φ_G by nonadiabatic processes to give a neutral Na ($^2S_{1/2}$) atom which was detected by a probe pulse with an established ultra fast detection technique.

8.3.2 Ionic Bond: Is the Bonding in HCl Covalent or Ionic?

In the previous section it was confirmed that the bond in NaCl in its ground electronic state is overwhelmingly ionic. But, what about the bond in HCl?

Since in aqueous solutions HCl molecules exist almost completely as H^+ (or more correctly, H_3O^+) and Cl^-, it might be considered that the bond of HCl in the ground electronic state was also ionic. This will be shown not to be the case through inspection of Fig. 8.5 below.

The vibrational spectrum of the gaseous HCl molecule has been thoroughly studied, and the equilibrium internuclear distance R_e, the dissociation

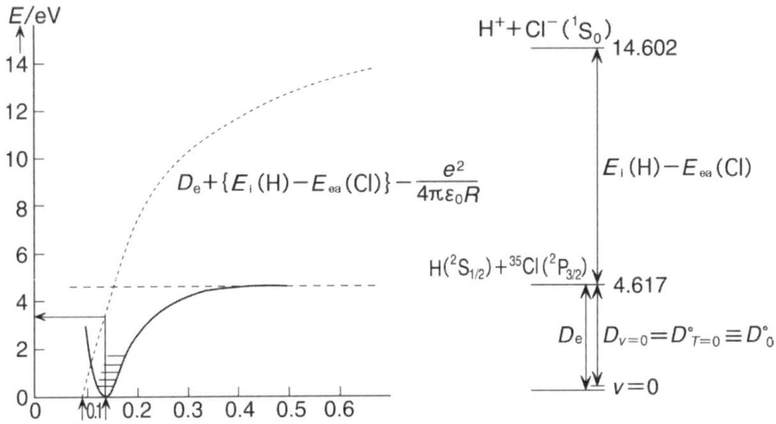

Fig. 8.5. Left: The solid curve represents the potential energy curve for the $H^{35}Cl$ molecule in the ground electronic state. The horizontal broken line indicates the limit of neutral dissociation at 4.617 eV. The broken curve shows a Coulombic potential asymptotically approaching the limit of ionic dissociation at 14.602 eV. The distances, $R_I = 0.0986$ nm and $R_e = 0.1274$ nm, are explained in the text. Right: The energy levels of the two limiting states. The difference between D_e (= 4.617 eV) and D°_0 corresponds to the zero-point energy, which is known to be 0.185 eV from spectroscopic data

energy D_e, and the zero-point energy are all well known, which are given in the caption of Fig. 8.5. Also, the ionization energy of the H atom and the electron affinity of the Cl atom have been determined to be $E_i(\mathrm{H}) = 13.598\,\mathrm{eV}$ and $E_{ea}(\mathrm{Cl}) = 3.613\,\mathrm{eV}$, respectively. Therefore, the intersection of the broken hyperbolic curve and the abscissa denoted by R_I in Fig. 8.5 can be calculated to be 0.0986 nm similar to the case in Fig. 8.4.

We notice that the energy given by the hyperbolic curve amounts to $+3.30\,\mathrm{eV}$ at $R_e = 0.1274\,\mathrm{nm}$. Such a substantial amount of energy means that the HCl molecule at its equilibrium internuclear distance cannot be approximated by a simple Coulombic pair of H^+ and Cl^- ions, contrary to the case of NaCl (cf. Fig. 8.4). Furthermore, various experimental data indicate that the ionic radius of Cl^- is 0.181 nm, definitely exceeding the equilibrium internuclear distance of $R_e = 0.1274\,\mathrm{nm}$. Thus, in no aspect can the bond of the HCl molecule be regarded as ionic; the character of the bond is mostly covalent as described by (8.69) and (8.70).

The fact that the HCl molecule dissociates into H^+ and Cl^- in aqueous solutions to give a strong acid has no direct relationship with the essential character of the bond in the HCl molecule itself, but rather the strong acidity is merely a result of a chemical reaction between HCl and liquid water molecules to yield H_3O^+ ions.

8.3.3 Coordinate Bond

The concept of the coordinate bond seems to have been introduced by A. Werner at the end of the nineteenth century. It was originally introduced to explain the fact that some inorganic ions attract more ions or atoms to form 'complex' compounds than expected from classical views of chemical bonding such as the interpretation in terms of the octet rule. Later on, the term 'coordinate bond' was also used to describe the bonding formed by a pair of some molecules consisting of lighter atoms in the second and third periods of the periodic table.

A typical coordinate bond is the N-B bond between the N atom of NH_3 and the B atom in BF_3 which is formed by the following reaction:

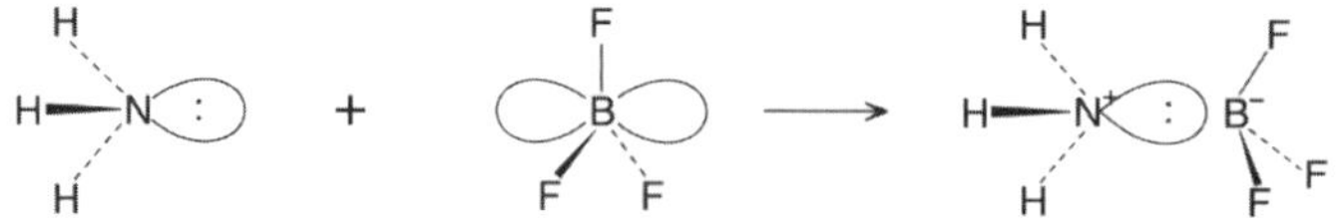

The two dots on the N atom of an NH_3 represent the two electrons occupying the $3a_1$ molecular orbital in Table 8.1 with their electron spins antiparallel. Such a notation is convenient to explain the octet rule proposed by G.N. Lewis in 1916, but is of course incapable of delineating the actual image of the electrons.

The lobe on the B atom in BF_3 represents schematically the vacant 2p orbital of the B atom perpendicular to the plane of the BF_3 molecule. The implication of the + and − signs on the product molecule is as follows. The net positive charge seen by the valence electrons of the N atom is +5 as a result of the screening of $Z = +7$ by the two inner shell electrons $1s^2$. Then, it may be interpreted that out of the five valence electrons $2s^2 2p^3$, three are associated with the three N-H bonds and one with the single N-B bond so that the net charge on the N atom may now be viewed as +1. The plus sign on the product molecule corresponds to this positive charge. Similarly, after the screening of $Z = +5$ in the B atom by the two inner shell electrons, the B atom is regarded as being surrounded by the three electrons associated with the three B-F bonds and another electron participating in the single N-B bond to make the overall charge −1. These +1 and −1 charges are called the formal charge. Obviously, these charges do not reflect correctly the actual electron distribution so that the charge is just 'formal' as it is called. The coordinate bond is sometimes called the semipolar bond. Such naming reflects an attempt to interpret the character of the bonding in terms of the classical electrostatic interaction without knowledge of quantum mechanical interpretations.

8.4 Some Topics and Prospects for the Study of Chemical Bonding

In this book I have put emphasis on the fundamental understanding of the chemical bond on the basis of the MO and VB theories. Due to various limitations, however, many important subjects are left untouched despite the fact that chemistry is a science directed towards the understanding of the diversity of matter. In the classic work by Pauling such subjects as the chemical bond involving d electrons, intermolecular bondings such as the hydrogen bond, and the metallic bond are fully treated along with a variety of discussions on the material properties associated with these various chemical bondings (L. Pauling, 1960).

Almost half a century has passed since that work during which our understanding of chemical bonds has been dramatically expanded thanks to the rapid progress in experimental techniques, such as laser spectroscopy, molecular beam techniques, mass-analytical spectroscopy, and computer analysis. Since it is impossible to cover all of this progress in this introductory text, I will briefly present a few recent developments as an excursion, while keeping to the level of this book.

8.4.1 Why does OF_6 not Exist, while SF_6 is a Stable Molecule?: The Discovery of CLi_6 as a Relevant Topic

SF_6 is a stable molecule with an octahedral structure belonging to the point group O_h. The central S atom has six equivalent S-F bonds, involving a total of twelve valence electrons. When the outermost electron shell of an atom in its ground state consists of only ns and np electrons, as in the case of an S atom with the configuration ($3s^2 3p^4$), the maximum number of valence electrons around the central atom is eight (cf. e.g., the N and B atoms in the complex of NH_3-BF_3 in Sect. 8.3.3), so that SF_6 apparently violates the octet rule and is sometimes called a hypervalent compound.

One of the explanations for this hypervalency was to propose that the S atom in the ground electronic configuration of ($\ldots 3s^2 3p^4$) could be promoted to an excited configuration of ($\ldots 3s 3p^3 3d^2$), which could then accept the six 2p electrons from the six F atoms. The same explanation can hardly be applied to the O atom with a similar configuration of ($\ldots 2s^2 2p^4$) because the energy necessary for the promotion to the required configuration of ($2s^2 2p^3 3d^2$) is much larger than in the case of the S atom (cf. Sect. 4.3.5). This is consistent with the absence of OF_6. In the following, however, it will be shown that resorting to 3d electrons to explain the stable existence of SF_6 is not a necessarily unique solution (B.M. Gimarc, 1979).

First, let us look at Fig. 8.6. The column for SF_6 in Fig. 8.6 demonstrates qualitatively that six occupied (one $1a_{1g}$, three $1t_{1u}$, and two $1e_g$) and four unoccupied (one $2a_{1g}$ and three $2t_{1u}$) orbitals can be constructed from a total of four 3s and 3p S atomic orbitals and of a total of six 2p F atomic orbitals. Each MO has 1 through 3 fold degeneracy according to the irreducible representations of the point group O_h as shown to the left of the figure, where the 2p orbitals of the F atoms are represented by circles for simplicity. The open and shaded circles differentiate the phase of the 2p orbitals of the F atoms from the phase of the overlapping 3s and 3p orbitals of the central S atom.

As is shown in the figure, the molecular orbitals $1a_{1g}$ and $1t_{1u}$ are of bonding character between S and F, while the $1e_g$ are of nonbonding (or lone pair) character. Likewise, $2a_{1g}$ and $2t_{1u}$ are anti-bonding.

The atomic radii of S and O atoms are estimated by experiments to be about 0.104 nm and 0.066 nm, respectively. Therefore, if the imaginary OF_6 molecule had an O_h structure similar to SF_6, the $1e_g$ orbital, which is nonbonding in the SF_6 molecule due to the sufficiently long S-F bonds, may become anti-bonding in the OF_6 molecule because the shorter O-F bond length may enhance the partial overlap of the adjacent 2p F atomic orbitals of opposite phases. If this is the case, the $1e_g$ orbital in the OF_6 molecule may be destabilized as shown in Fig. 8.6 and the orbital energy may become positive crossing over the level of $\varepsilon = 0$, called the Fermi level. Then, not all the twelve valence electrons of the OF_6 molecule may be attracted in the vicinity of the central O atom and stable formation of the molecule may fail. The intersection of the broken lines between the columns of the SF_6 and

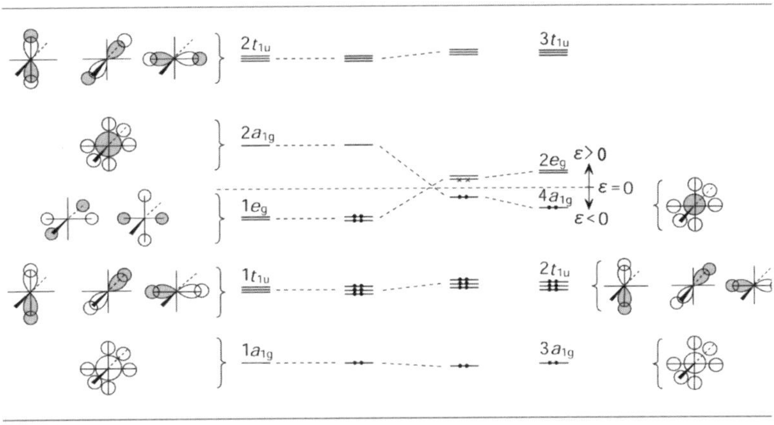

Fig. 8.6. Qualitative correlation diagram among the MO energies of SF_6, OF_6, and CLi_6

OF_6 molecules in Fig. 8.6 indicate such an interchange of the stability of the electrons, which is conceivable when we consider that the distance between S and F in SF_6 is noticeably longer than the distance between O and F in OF_6.

So far, the argument is totally qualitative. However, there is a supporting experimental finding in connection with the argument here, which is the result found for the CLi_6 molecule (H. Kudo and K. Yokoyama, 1996): the formation of the CLi_6 molecule was confirmed through mass-spectrometry of a gaseous sample obtained by heating the known compound Li-C_2-Li (lithium carbide). Molecular orbital analyses of the CLi_6 molecule reveal that the orbitals below the Fermi level are ordered as shown in Fig. 8.6. The figure shows that the $1e_g$ orbital, which is nonbonding in the SF_6 molecule, becomes anti-bonding in the CLi_6 molecule as surmised above for the OF_6 molecule. On the other hand, the $2a_{1g}$ orbital, which is anti-bonding in the SF_6 molecule, was found to be stabilized in the CLi_6 molecule and is located below the Fermi level. As a result, all the ten valence electrons of the CLi_6 molecule are predicted to be stably bound around the central C atom. According to that MO calculation, the equilibrium C-Li bond length was estimated to be 0.206 nm, which is fairly long for a purely covalent molecule. The six Li atoms surrounding the central C atom may be regarded as forming a kind of a metallic shell which is positively charged as a whole. Corresponding to this positive charge, an equal amount of negative charge is induced in the central C atom so that the electrostatic attractive interaction is regarded as contributing significantly to the formation of the CLi_6 molecule.

Thus, within the framework of the MO calculation involving only the ns and np atomic orbitals, there is no need to invoke the hypervalent bond. However, as higher accuracy in the MO calculation is demanded, it will become necessary to include more and more basis functions as stated in Sect. 8.1. As such basis functions, d-type AOs may also be required, in which case there is little significance in classifying the type of chemical bond as valent or hypervalent.

Retrospectively speaking, the term 'valence bond' or 'covalent bond' was introduced through the naive intuition of chemists, much like the concept of Lewis' octet rule. In this context, it bears repeating that the bond in HCl discussed in Sect. 8.3.2 can be attributed to neither a simple ionic nor a valence bond. Similar remarks are appropriate to the coordinate bond in Sect. 8.3.3.

In chemistry, newly discovered chemical species, phenomena, subjects, and concepts are first given names to describe their superficial features. But, as understanding is deepened, the initial nomenclature often becomes restrictive. An example is the case of the hybridized orbital discussed in Sect. 8.2.5.

By the same token, the two apparently different approaches used by the MO and VB theories may be merged into one and the same tool to describe the reality of molecules as the basis sets and configurations are expanded in both theories.

8.4.2 Molecules Comprising Rare Gas Atoms

For the discussion in this section we first take a look at the (first) ionization energy of rare gas atoms and some main group elements in Table 8.3.

The ionization energy of the rare gas atoms decreases with the atomic number due to the increase in the shielding effect for the outermost electrons. With the exception of the first two members, He and Ne, the ionization energies are within the range of about $13 \pm 3\,\text{eV}$, which is not so different from the ionization energies of the atoms of the main group elements in the table. Since all the rare gas atoms have closed electronic shell structures, it is implausible for them to accept an electron from other atoms. However, as for the ionization, the similarity of the ionization energies to those of non-rare gas atoms in Table 8.3, except for He and Ne, suggests that the heavier rare gas atoms have the potential to donate one or more electrons to non-rare gas atoms.

Table 8.3. The first ionization energies of the rare gas atoms and atoms of some of the main group elements in units of eV

He	Ne	Ar	Kr	Xe	Rn
24.587	21.565	15.760	14.000	12.130	10.749
H	B	C	N	O	F
13.598	8.298	11.260	14.534	13.618	17.422

Pauling (1933) predicted the possibility of chemical bond formation for the heavier rare gas atoms. This prediction, in fact, was substantiated some 30 years later (N. Bartlett, 1962). Bartlett, after a successful synthesis of the complex compound O_2PtF_6, tried to synthesize $XePtF_6$ by noticing the proximity of the ionization energies of O_2 (12.70 eV) and of Xe (12.13 eV). The experiment was successful and a number of Xe containing compounds such as XeF_4 and XeF_2 were obtained in the form of crystal (H.H. Classen, 1962).

Let us look into the nature of the chemical bond in these Xe containing compounds, first, from the MO theory perspective. In order to emphasize the essential feature of the bond we will simplify the system drastically, i.e., in the case of the linear molecule of XeF_2, only one 5p orbital of the Xe atom and one 2p orbital of each F atom are taken into account.

The LCAO-MOs are schematically shown in Fig. 8.7(b) where the coefficients of the linear combinations are shown qualitatively. Figure 8.7(a) demonstrates semiquantitatively the splitting of the orbital energies for the σ and π types. Due to the difference in the interaction between the atomic orbitals, the magnitude of the splitting in the σ type is larger than that in the π type. The two energies indicated by the vertical arrows correspond to the ionization energies of the 5p orbital of the Xe atom and the 2p orbital of the F atom which, are approximately equal to the orbital energies in accordance with Koopmans' theorem.

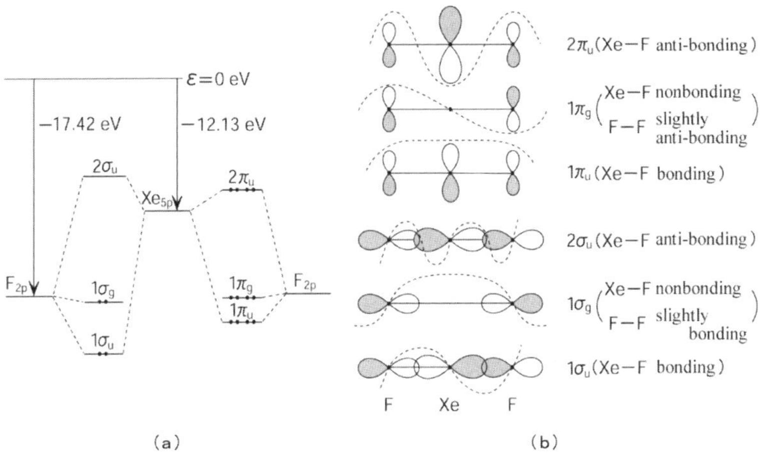

Fig. 8.7. Schematic representation of the MOs of XeF_2
(**a**) Semiquantitative splittings of the σ type orbitals (left) and of the π type orbitals (right). The energies attached to the vertical arrows represent the first ionization energies of the Xe and F atoms.
(**b**) Qualitative patterns of the six MOs

As to the notations σ and π in the figure, refer to the explanation on the molecular orbitals of HF in Sect. 8.2.1. The subscripts g (German 'gerade') and u (German 'ungerade') mean that the orbital phase does not, and does change under inversion in the center of the molecule, respectively.

In the simplified model, the total of sixteen electrons, i.e., six from the Xe 5p shell and ten from the two F 2p shells, are assumed to occupy the two σ type orbitals and the three π type orbitals, as shown in Fig. 8.7(a). When the molecular axis is taken as parallel to the x-axis, each π type orbital consisting of the perpendicular degenerate p_y and p_z AOs accommodates four electrons, as in Fig. 8.7(a). These electrons do not contribute significantly to the stabilization of the molecule as a whole. On the other hand, the occupied σ type orbitals are composed of two $5p_x$ electrons from the Xe and four $2p_x$ electrons from the two F atoms. For this reason the bond may be called a three-center-four-electron bond.

From the perspective of VB theory, on the other hand, the molecule XeF_2 can be regarded as resulting from the superposition of configurations F^-Xe^+-F and F-Xe^+F^-. Here, the usage of the term 'superposition' is consistent with the usage of the term made heretofore in this book, however Pauling called it the 'resonance' of the two configurations. In both configurations shown above the electronic shell structure of Xe^+ is $5p^5$, similar to that of the isoelectronic I atom, while F^- has a shell structure of $2p^6$, similar to that of the isoelectronic Ne atom. The open shell structured Xe^+ and F are to be bonded covalently. On the other hand, the pair Xe^+ and F^- are attracted by Coulombic interaction. In addition to these conditions favorable for bond formation, the whole molecule is stabilized by the exchange interaction whose importance has been repeatedly emphasized since Chap. 6.

The study of molecules containing rare gas atoms became dormant for a period until recent developments; e.g., it has been determined that not only Xe, but also Kr and Ar can yield fairly varied molecules. In addition, the chemical bonding between rare gas atoms and the H atom has been confirmed spectroscopically. Such molecules include H-Rg-H, H-Rg-X, and H-Rg-X-Y, where Rg stands for rare gas atoms and X and Y are non-hydrogenic atoms (M. Räsänen, 2000).

The new molecule of HArF exhibits infrared absorption spectra typical for a three-atom molecule. According to the experiment cited, the frequencies of the Ar-H stretching, Ar-F stretching, and H-Ar-F bending motions are determined to be $1969.5\,\mathrm{cm}^{-1}$, $687.0\,\mathrm{cm}^{-1}$, and $435.7\,\mathrm{cm}^{-1}$, respectively, which are fairly close to the computational results of $2097\,\mathrm{cm}^{-1}$, $729\,\mathrm{cm}^{-1}$, and $480\,\mathrm{cm}^{-1}$, respectively. Computations predict that the bond lengths of H-Ar and Ar-F are 0.133 nm and 0.197 nm, respectively. These lengths compare favorably with the known lengths of 0.127 nm for H-Cl and of 0.160 nm for Cl-F, which are analogous to H-Ar and Ar-F, respectively. As for the character of the bonds in HArF, the electrical charges on the Ar and F atoms are computed to be $+0.54$ and -0.76, respectively, which means that the bond is mainly ionic with a non-negligible covalent bond contribution. The cited work is educational in the sense that earlier preoccupation with the inertness of the noble gasses is shown to be groundless.

In deference to the publisher's request to discuss the prospects of future studies in chemical bonding, I will present rather capriciously a few topics that might be pursued in the future. As such, I consider the in depth study of intermolecular interactions important because the energy state functions of the molecules so far discussed are determined on the basis of the criteria of the variational minimum of the total energy of individual molecules. Such criteria do not necessarily guarantee the reliability of the description of the energy state functions for a point remote from the central part of the molecules. Subtle details of the state function at such a distance may play a crucial role in determining intermolecular interactions, which may play a significant role in, e.g., biological systems. After all, we need more information on the energy state function over a wide region of space, which will require our continuing effort.

When it comes to intermolecular interactions, the hydrogen bond and bonding by the van der Waals interaction are traditionally familiar. Molecular aggregates formed by these bondings are called clusters in general, which are currently being studied actively both experimentally and theoretically. The motivations for such studies are multiple, but one of the main motivations is to disclose new information on clusters as an intermediate phase between the gaseous and condensed phases and to uncover any useful potential applications of that new phase.

As a glance at the recent progress in the research activity on clusters, I will introduce a recent study of the hydrogen bond in the next section. The study of the hydrogen bond has a long history, but it is still growing because of its significance, especially in biological systems. Such a study is very basic but is important in the steady advancement towards the clarification of subtle intermolecular interactions.

8.4.3 An Example of a Recent Study of the Hydrogen Bond

The linear molecule of HCN has an electric dipole moment as high as 3.05 D (1D (Debye) $\cong 3.335 \times 10^{-30}$ C m). On account of this high dipole moment a hydrogen bond is formed between two HCN molecules. If the intermolecular interaction was solely due to the electric dipole moment, the dimer of HCN in the gas phase would be in a head-to-tail collinear conformation, H–C≡N...H–C≡N according to electrostatic theory. However, in the gas phase the two HCN molecules can rotate almost freely depending on the ambient temperature, and obtain a certain rotational energy. At room temperatures, this rotational energy of each HCN molecule surpasses the electrostatic stabilization energy of the linear dimeric molecule so that the linear conformation of the dimer is lost.

At very low temperatures, however, the rotational energy is suppressed and the linear structure of the dimer is regained. This linear dimer can align a third HCN molecule, forming a linear trimer. According to experiment, it has been confirmed that about ten HCN molecules continue to align in a straight chain (R.E. Miller, 1992).

This alignment of the HCN molecule is an example of the process generally called self-assembly. In supramolecular chemistry, a new branch of modern chemistry, the self-assembling process is of central interest. For such a new field of chemistry to expand, steady fundamental studies are indispensable, needless to say.

8.4.4 Interstellar Molecules

In Chaps. 7 and 8, we were mostly concerned with molecules with closed electronic shell structures. This is because most molecules have such shell structures and because theoretical treatments of molecules with open electronic shell structures is much more complicated (G. Herzberg, 1971).

Most molecules with open electronic shell structures are chemically reactive so as to be short-lived under usual conditions. The high reactivity, however, means that the open shell molecules are an interesting subject from the viewpoint of chemical reactions, for example. These chemically reactive species can be observed by means of various detection techniques in the laboratory (T. Shida, 1991,1994, for example). Some of them can be detected in the interstellar space also where the frequency of bimolecular collisions is drastically suppressed. As a matter of fact, a number of new open shell molecules, along with closed shell molecules, are being discovered in interstellar space and identified unambiguously by spectroscopy, the techniques of which are continuously being developed. Since the first discovery of a so-called interstellar molecule in 1937 (G. Herzberg, 1987), more than one hundred interstellar molecules are now known. Examples of interstellar molecules having the chemical formula C_nH are shown in Fig. 8.8 (E. Hirota, 1992).

In Fig. 8.8, the symbols representing schematically the hybrid bonds have been confirmed by quantum chemical computations on these molecules.

Formerly, astronomers seldom paid attention to such things as molecules in space. Nowadays, however, many molecules provide valuable information

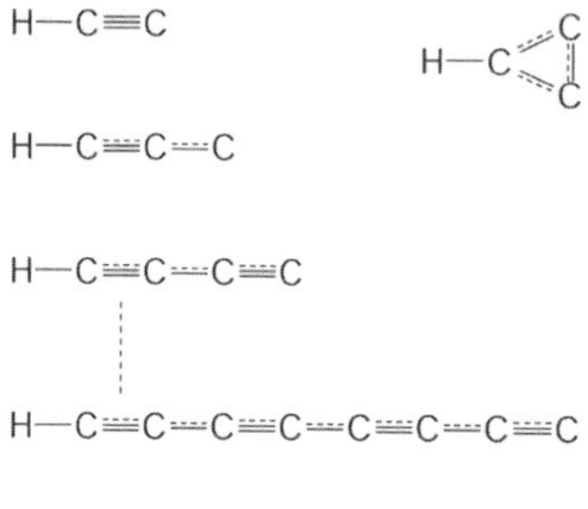

Fig. 8.8. Some examples of the interstellar molecule

to astronomy. A most impressive achievement is the discovery of the simplest polyatomic molecule H_3^+, which turns out to have very basic significance in cosmochemistry (T. Oka, 2000). These interstellar molecules so far have little to do with our daily life. However, when we look back on history, it often happens that ostensibly unrelated subjects suddenly come together as the basis for profound new discoveries. As such an example, one may consider the discovery of fullerene C_{60}, stimulated by radioastronomical observations. This is a good lesson which teaches us the importance of interdisciplinary studies.

As a concluding remark, I would like to emphasize that the study and research of basic science is of utmost importance for us and must be supported by all means, all the more so as some people tend to expect quick success in 'useful' sciences only.

References

It is proper and polite to cite every reference individually. However, this manner was not followed for the reasons stated in the preface. Instead, the references that I have consulted will be listed here chapter by chapter, to express my thanks to the authors.

Some of the references, however, may be beyond the elementary level, and some are out of print. Nevertheless, I have tried as much as possible to extract, as simply as possible, the original messages that I think important. I hope the readers themselves will consult the original references at some time in the future.

Some recent publications are added in the English version.

Chap. 1

A.J. Ihde, *The Development of Modern Chemistry*, Harper & Row, 1964.

S. Weinberg, *The Discovery of Subatomic Particles*, Scientific American, 1982.

H. Nagaoka, Philosophical Magazine, **7** (1904) p. 445. (Incidentally, on p. 237 of the same volume of this journal one can find the often cited article by J.J. Thomson on the so-called pudding model of atoms).

J. Avery, *The Quantum Theory of Atoms*, Molecules, and Photons, McGraw-Hill, 1972.

D. Bohm, *Quantum Theory*, Prentice-Hall, 1951.

T.W. Hänsch and A.L. Schawlow, *The Spectrum of Atomic Hydrogen*, Scientific American, 1979.

J.S. Rigden, *Hydrogen: The Essential Element*, Harvard University Press, 2002.

Chap. 2

This chapter is devoted to excerpt the essentials of quantum mechanics. There are countless textbooks on quantum mechanics. As some of the recently published textbooks, I will list rather capriciously the following monographs. Most of them are intended for advanced undergraduate and the first-year graduate students of physics but I think they are not much beyond the reach for motivated chemistry students too.

W. Gleiner, *Quantum Mechanics: An Introduction*, Springer, 1989.

R. Shankar, *Principles of Quantum Mechanics*, 2nd ed.,Plenum, 1994.

S. Gasiorowicz, *Quantum Physics*, 2nd ed., John Wiley & Sons, 1995.

D.D. Fitts, *Principles of Quantum Mechanics*, Cambridge University Press, 1999.

M.D. Fayer, *Elements of Quantum Mechanics*, Oxford University Press, 2001.
N. Zettili, *Quantum Mechanics*, John Wiley & Sons, 2001.
F.S. Levin, *An Introduction to Quantum Theory*, Cambridge University Press, 2002.
L. Marchildon, *Quantum Mechanics*, Springer, 2002.
F. Schwabl, *Quantum Mechanics*, 3rd ed., Springer, 2002.

The following five are the references cited or consulted in Chap. 2.

M. Born, *Atomic Physics*, 8th ed., Dover, 1989.
J.D. Macomber, *The Dynamics of Spectroscopic Transitions*, John Wiley & Sons, 1976.
A.P. French and E.F. Taylor, *MIT Introductory Physics Series, An Introduction to Quantum Mechanics*, 1st ed., Norton, 1978.
S. Brandt and H.D. Dahmen, *The Picture Book of Quantum Mechanics*, 3rd ed., Springer, 2001.
C.S. Johnson, Jr. and L.G. Pedersen, *Problems and Solutions in Quantum Chemistry and Physics*, Addison-Wesley, 1974; the paperback is reprinted from Dover since 1986.

Chap. 3

S. Flügge, *Practical Quantum Mechanics*, Springer, 1974.
E.U. Condon and G.H. Shortley, *The Theory of Atomic Spectra*, Cambridge University Press, 1935; the paperback is reprinted from Cambridge University Press since 1970.
A.C.G. Mitchell and Z.W. Zemansky, *Resonance Radiation and Excited Atoms*, Cambridge University Press, 1934.
R. Shankar, *Principles of Quantum Mechanics*, 2nd ed., Plenum Press, 1994.
M. Weissbluth, *Atoms and Molecules*, Academic Press, 1978.
E. Steiner, *The Determination and Interpretation of Molecular Wave Functions*, Cambridge University Press, 1976.

Chap. 4

J.A. Pople and D.L. Beveridge, *Approximate Molecular Orbital Theory*, McGraw-Hill, 1977.
H. Friedrich, *Theoretical Atomic Physics*, 2nd ed., Springer, 1998.
S. Bashkin and J.O. Stoner, Jr., *Atomic Energy Levels and Grotorian Diagrams*, North-Holland, 1975.
A.R. Edmonds, *Angular Momentum in Quantum Mechanics*, Princeton University Press, 1960.
C.E. Moore, *Tables of Spectra of Hydrogen, Carbon, Nitrogen, and Oxygen Atoms and Ions*, CRC Press, 1993.
I.N. Levine, *Quantum Chemistry*, 5th ed., Prentice Hall International, 2000.
J.I. Steinfeld, *Molecules and Radiation*, Harper & Row, 1974.
J.H. Van Vleck, Journal of Chemical Physics, **1** (1933) p. 177 and p. 219; *ibid.*, **2** (1934) p. 20 and p. 296.
L. Pauling, Journal of the American Chemical Society, **53** (1931) p. 1367.

M. Karplus and R.N. Porter, *Atoms and Molecules: An Introduction for Students of Physical Chemistry*, W.A. Benjamin, 1970.
H. Eyring (ed)., *Physical Chemistry, An Advanced Treatise Vol. 5 Valency*, Academic Press, 1970.
C.F. Fischer, *The Hartree-Fock Method for Atoms, A Numerical Approach*, John Wiley & Sons, 1977.
V.M.S. Gil, *Orbitals in Chemistry*, Cambridge University Press, 2000; the book contains a number of educational articles cited from many volumes of Journal of Chemical Education and is useful for Chap. 8 also.

Chap. 5

F. Jensen, *Introduction to Computational Chemistry*, John Wiley & Sons, 1999.
W.H. Flygare, *Molecular Structure and Dynamics*, Prentice-Hall, 1978.

Chap. 6

G. Herzberg, *Molecular Spectra and Molecular Structure, I. Spectra of Diatomic Molecules*, D. van Nostrand, 1945; the book is reprinted by R.E. Krieger in 1989.
Sin-itiro Tomonaga, *Spin wa meguru* (in Japanese), Chuokoron-sha, 1974; (English translation: T. Oka, *The Story of Spin*, The University of Chicago Press, 1997).
R.P. Feynman, R.B. Leighton, and M. Sands, *The Feynman Lectures on Physics, Vol. 5, Quantum Mechanics*, Addison-Wesley, 1965.
A. Carrington, Journal of Chemical Society, Faraday Transactions, **91** (1995) p. 1887.
J.C. Slater, *The Calculation of Molecular Orbitals*, John Wiley & Sons, 1979.
J. Posthumus, *Molecules and Clusters in Intense Laser Fields*, Cambridge University Press, 2001.

Chap. 7

E.E. Witmer, Physical Review, **28** (1926) p. 1223.
H. Beutler, Zeitschrift für physikalische Chemie, **B27** (1934) p. 315.
W. Heitler and F. London, Zeitschrift für Physik, **44** (1927) p. 455.
Y. Sugiura, Zeitschrift für Physik, **45** (1927) p. 484.
A. Kojima and K. Higasi (eds.), *Theories of Chemical Bond I. (Classics of Chemistry Series 1)*, (in Japanese), The Chemical Society of Japan, 1975.
M. Karplus and R.N. Porter, *Atoms and Molecules: An Introduction for Students of Physical Chemistry*, W.A. Benjamin, 1970.
H.M. James and A.S. Coolidge, Journal of Chemical Physics, **1** (1933) p. 825.
W. Kolos and L. Wolniewicz, Journal of Chemical Physics, **43** (1965) p. 2429.
G. Herzberg, Journal of Molecular Spectroscopy, **33** (1970) p. 147.
I. Dabrowski, Canadian Journal of Physics, **62** (1984) p. 1639.
W. Kolos and L. Wolniewicz, Chemical Physics Letters, **24** (1974) p. 457.
D.A. McQuarrie and J.D. Simon, *Physical Chemistry, A Molecular Approach*, University Science Books, 1997.

Chap. 8

W.J. Hehre, L. Radom, P.v.R. Schleyer, J.A. Pople, *Ab Initio Molecular Orbital Theory*, John Wiley & Sons, 1986.

R.G. Parr and W. Yang, *Density-Functional Theory of Atoms and Molecules*, Oxford University Press, 1989.

J.B. Foresman and Æ. Frisch, *Exploring Chemistry with Electronic Structure Methods*, 2nd ed., Gaussian, Inc., 1996.

A. Szabo and N.S. Ostrand, *Modern Quantum Chemistry, Introduction to Advanced Electronic Structure Theory*, Dover, 1996.

F. Jensen, *Introduction to Computational Chemistry*, John Wiley & Sons, 1999.

I.N. Levine, *Quantum Chemistry*, 5th ed., Prentice Hall International, 2000.

C.C. Roothaan, Review of Modern Physics, **23** (1951) p. 66.

S. Huzinaga, *Molecular Orbital Methods* (in Japanese), Iwanami, 1980.

B.J. Ransil, Review of Modern Physics, **32** (1960) p. 245.

J.A. Pople and D.L. Beveridge, *Approximate Molecular Orbital Theory*, McGraw-Hill, 1977.

S.F. Boys, Proceedings of the Royal Society (London), **A200** (1950) p. 542.

R.M. Pitzer and D.P. Merrifield, Journal of Chemical Physics, **52** (1970) p. 4782.

R.M. Pitzer, Journal of Chemical Physics, **46** (1967) p. 4871.

W.E. Palke and W.N. Lipscomb, Journal of American Chemical Society, **88** (1966) p. 2384.

J.W. Rabalais, *Principles of Ultraviolet Photoelectron Spectroscopy*, John Wiley & Sons, 1977.

C.F. Fischer, *The Hartree-Fock Method for Atoms, A Numerical Approach*, John Wiley & Sons, 1977.

G. Herzberg, *Molecular Spectra and Molecular Structure, III. Electronic Spectra and Electronic Structure of Polyatomic Molecules*, D.van Nostrand, 1967; the book is reprinted by R.E. Krieger in 1991.

Foster and S. Boys, Review of Modern Physics, **32** (1960) p. 300.

C. Edmiston and K. Ruedenberg, Review of Modern Physics, **35** (1963) p. 457.

L. Pauling, Journal of the American Chemical Society, **53** (1931) p. 1367.

E. Wigner and E.E. Witmer, Zeitschrift für Physik, **51** (1928) p. 859.

T.S. Rose, M.J. Rosker and A.H. Zewail, Journal of Chemical Physics, **88** (1988) p. 6672.

L. Pauling, *The Nature of Chemical Bond and the Structure of Molecules and Crystals*, Cornell University Press, 1960.

B.M. Gimarc, *Molecular Structure and Bonding*, Academic Press, 1979.

H. Kudo and K. Yokoyama, Bulletin of the Chemical Society of Japan, **69** (1996) p. 1459.

L. Pauling, Journal of the American Chemical Society, **55** (1933) p. 1895.

N. Bartlett, Proceedings of the Chemical Society, (1962) p. 218.

H.H. Claassen, H. Selig and J.G. Malm, Journal of the American Chemical Society, **84** (1962) p. 3593.

M. Räsänen et al., Nature, **406** (2000) p. 874.

R.E. Miller et al.,Physical Review Letters, **68** (1992) p. 1303 and succeeding papers.

G. Herzberg, *The Spectra and Structures of Simple Free Radicals. An Introduction to Molecular Spectroscopy*, Dover, 1971.

T. Shida, Annual Review of Physical Chemistry, **42** (1991) p. 55.
T. Shida, T. Kato, and T. Momose, *Studies in Physical and Theoretical Chemistry, Vol. 82, Dynamics of Excited Molecules*, K. Kuchitsu (ed)., p. 559, Elsevier, 1994.
G. Herzberg, Science, **238** (1987) p. 1341.
E. Hirota, Chemical Reviews, **92** (1992) p. 141.
T. Oka, Phil. Trans. R. Soc. Lond. A **358** (2000) p. 2363.

Index

Printing: Saladruck, Berlin
Binding: Stein+Lehmann, Berlin